Advances in Geological Science

Studies in the twentieth century uncovered groundbreaking facts in geophysics and produced a radically new picture of the Earth's history. However, in some respects it also created more puzzles for the research community of the twenty-first century to tackle. This book series aims to present the state of the art of contemporary geological studies and offers the opportunity to discuss major open problems in geosciences and their phenomena. The main focus is on physical geological features such as geomorphology, petrology, sedimentology, geotectonics, volcanology, seismology, glaciology, and their environmental impacts. The monographs in the series, including multi-authored volumes, will examine prominent features of past events up to their current status, and possibly forecast some aspects of the foreseeable future. The guiding principle is that understanding the fundamentals and applied methodology of overlapping fields will be key to paving the way for the next generation.

More information about this series at https://link.springer.com/bookseries/11723

Yasuhiro Kumahara · Heitaro Kaneda ·
Hiroyuki Tsutsumi
Editors

Surface Ruptures Associated with the 2016 Kumamoto Earthquake Sequence in Southwest Japan

Springer

Editors
Yasuhiro Kumahara
Graduate School of Humanities and Social Sciences
Hiroshima University
Hiroshima, Japan

Heitaro Kaneda
Department of Civil and Environmental Engineering
Chuo University
Tokyo, Japan

Hiroyuki Tsutsumi
Department of Environmental Systems Science
Doshisha University
Kyoto, Japan

ISSN 2524-3829 ISSN 2524-3837 (electronic)
Advances in Geological Science
ISBN 978-981-19-1149-1 ISBN 978-981-19-1150-7 (eBook)
https://doi.org/10.1007/978-981-19-1150-7

This Springer imprint is published by the registered company Springer Nature Singapore Pte Ltd.
The registered company address is: 152 Beach Road, #21-01/04 Gateway East, Singapore 189721, Singapore

Preface

We are writing this Preface in April 2021, five years after the 2016 Kumamoto earthquake sequence in Kyushu, southwest Japan. A series of seismic activities began on the night of 14 April 2016 with an Mw 6.2 event followed by a larger Mw 7.0 event a full day later. This earthquake sequence was the first since the Japanese instrumental seismic observation began in 1885, that two earthquakes with the Japan Meteorological Agency (JMA) seismic intensity maximum scale 7 were recorded at the same location. The Kumamoto earthquake sequence killed more than 200 people and destroyed totally and partially more than 40,000 houses and was the deadliest inland earthquake in Japan since the 1995 Kobe earthquake. Immediately after the earthquake, our group, composed of more than 25 researchers from Japanese universities and research institutes, started to map the coseismic surface ruptures. Because the surface breaks were long and widespread, it took us a couple of years to complete the mapping activities. These efforts clarified that this earthquake was caused by the reactivation of the previously mapped Futagawa and surrounding active faults and revealed the whole picture of the surface ruptures associated with the earthquake sequence. While a few research groups promptly published their mapping results in research journals and analyzed the general characteristics of the surface ruptures, this monograph aims to provide a complete surface faulting record of the earthquake sequence, featuring large-scale maps of all the surface ruptures, photographs taken immediately after the earthquake, detailed explanations in English, and digital data of the surface rupture traces and slip measurement points. We hope that this monograph will contribute to a better understanding of the 2016 Kumamoto earthquake sequence and further promotion of earthquake sciences and earthquake hazard mitigation studies.

We owe many people in the production of this monograph. During our field surveys, residents allowed us to access their properties and provided eyewitness accounts even when they were devastated by the damages from the earthquake sequence. Researchers from various universities and institutes kindly shared their observations and ideas, which greatly facilitated our field mapping. Many residents allowed us to conduct geological excavations on their properties. The Geospatial Information Authority of Japan provided us post-earthquake aerial photographs and digital elevation model data that were utilized in many figures in this monograph. The JSPS grants supported some of our post-earthquake researches. Finally, but not least, we express our condolences to the people who lost their lives by the earthquake. Although many infrastructures have been reconstructed and repaired, the recovery of the affected areas is still underway. We would sincerely hope that this monograph also plays a role in building more earthquake-resilient societies.

Hiroshima, Japan — Yasuhiro Kumahara
Tokyo, Japan — Heitaro Kaneda
Kyoto, Japan — Hiroyuki Tsutsumi
April 2021

Contents

Part I

Tectonic Setting of the Epicentral Area

Geomorphology and Geology

1

Hiroyuki Tsutsumi, Heitaro Kaneda, and Yasuhiro Kumahara

Abstract

This chapter introduces the geomorphic and geologic settings of the epicentral area of the 2016 Kumamoto earthquake sequence. The earthquake ruptured a part of the southern boundary faults of the Beppu-Shimabara graben zone, a tectonic depression characterized by active volcanism and normal faulting in central Kyushu. Main geomorphic features and geologic units in and around the Kumamoto Plain and Aso Volcano are described.

Keywords

2016 Kumamoto earthquake • Beppu-Shimabara graben zone • Aso Volcano • Normal fault • Kumamoto Plain • Median Tectonic Line

Supplementary Information
The online version contains supplementary material available at (https://doi.org/10.1007/978-981-19-1150-7_1).

Note
The supplementary material of this book contains geographical information system (GIS) and spreadsheet files of the surface ruptures associated with the mainshock of the 2016 Kumamoto earthquake sequence

H. Tsutsumi (✉)
Department of Environmental Systems Science,
Faculty of Science and Engineering, Doshisha University,
1-3, Tatara-Miyakodani, Kyotanabe, Kyoto, 610-0394, Japan
e-mail: htsutsum@mail.doshisha.ac.jp

H. Kaneda
Department of Civil and Environmental Engineering, Chuo University, 1-13-27, Kasuga, Bunkyo-ku, Tokyo, 112-8551, Japan
e-mail: kaneda.00c@g.chuo-u.ac.jp

Y. Kumahara
Graduate School of Humanities and Social Sciences,
Hiroshima University, 1-1-1, Kagamiyama, Higashi-Hiroshima, Hiroshima, 739-8524, Japan
e-mail: kumakuma@hiroshima-u.ac.jp

The epicentral area of the 2016 Kumamoto earthquake sequence is located at the southern margin of a volcano-tectonic depression called "the Beppu-Shimabara graben zone (BSGZ)" that traverses central Kyushu Island from Beppu Bay to Tachibana Bay in a NE-SW direction (Figs. 1.1 and 1.2; Matsumoto 1979). There are a series of active volcanos, the Tsurumi, Yufu, Kuju, Aso, and Unzen volcanos from east to west, and a dense network of east-trending active normal faults in the BSGZ. This zone, especially the eastern part, is characterized by negative Bouguer gravity anomalies (Fig. 1.3), resulting from the subsidence of basement rocks and the thick accumulation of low-density volcanic deposits. Repeated triangulation and leveling surveys in the past ~100 years indicate that the zone has extended in a north–south direction at a rate of 1–2 cm/year, and the axial area has subsided at a rate of 2–3 cm/year (Tada 1993).

The dense distribution of active normal faults in central Kyushu is unique in the Japanese islands located at convergent plate boundaries. Therefore, the origin of the extensional stress field in the BSGZ has been a topic of great interest and investigated by numerous studies. There are two representative hypotheses. Eguchi and Uyeda (1983) proposed that the BSGZ is the northeastern onshore extension of the Okinawa Trough, the active back-arc rift zone of the Ryukyu arc. They interpreted that the graben structures and volcanism in the BSGZ are related to the back-arc rifting. In contrast, Kamata (1989) proposed that the formation of the graben structures is associated with the right-lateral slip on the Oita-Kumamoto Tectonic Line, a western extension of the Median Tectonic Line active fault zone, at the southern margin of the BSGZ (Fig. 1.2). The oblique subduction of the Philippine Sea Plate beneath the Eurasian Plate causes the right-lateral slip on the tectonic line. The eastern translation of the crustal block north of the Oita-Kumamoto Tectonic Line resulted in the subsidence of the trailing edge of the block. This zone of subsidence is traversed obliquely by the volcanic front associated with the Ryukyu trench.

Y. Kumahara et al. (eds.), *Surface Ruptures Associated with the 2016 Kumamoto Earthquake Sequence in Southwest Japan*, Advances in Geological Science, https://doi.org/10.1007/978-981-19-1150-7_1

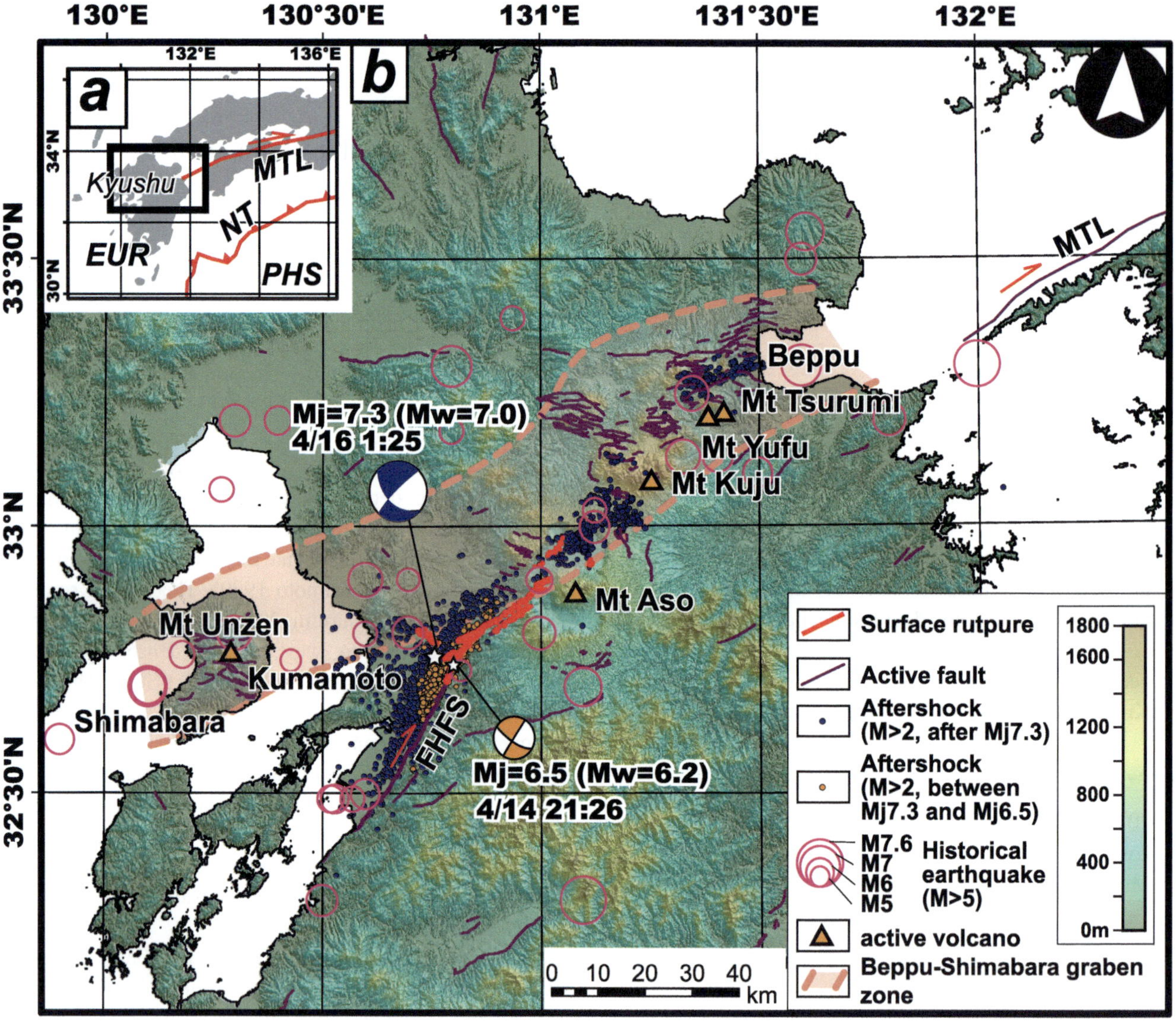

Fig. 1.1 Tectonic setting of central Kyushu Island and the Beppu-Shimabara graben zone. **a** Plate tectonic setting of Kyushu. EUR: Eurasian Plate, PHS: Philippine Sea Plate, NT: Nankai Trough, MTL: Median Tectonic Line. **b** Tectonic setting of the epicentral area of the 2016 Kumamoto earthquake sequence. FHFS: Futagawa-Hinagu fault system

Even in this hypothesis, the Unzen Volcano is interpreted to be related to the Okinawa Trough. The Futagawa fault, which ruptured during the Kumamoto earthquake sequence, is considered a segment of the Oita-Kumamoto Tectonic Line (Figs. 1.1 and 1.2).

The surface ruptures of the 2016 Kumamoto earthquake sequence appeared in and around the Kumamoto Plain and Aso caldera. The Kumamoto Plain and its northern hills are bounded on the north by the Chikuhi and Tsue Mountains (Fig. 1.2). The Chikuhi Mountains are underlain primarily by the Sangun metamorphic rocks of late Triassic–late Jurassic age. In the Tsue Mountains, volcanic rocks of Pliocene–middle Pleistocene age overlie the basement rocks. The Kumamoto Plain is bounded on the south by the Mashiki Mountains, composed of the early Cretaceous Higo metamorphic rocks and Permian–Cretaceous sedimentary rocks. The Kinpozan Mountain, located northwest of the Kumamoto Plain, is an early–middle Pleistocene volcano with a caldera 2.5 km in diameter. Several east-trending normal faults cut the volcanic slope (Machida et al. 2001).

The Aso Volcano, located northeast of the Kumamoto Plain, is one of the largest caldera volcanos on the Japanese islands (Geological Society of Japan 2010). The caldera is ~25 km long in the north–south direction and ~18 km

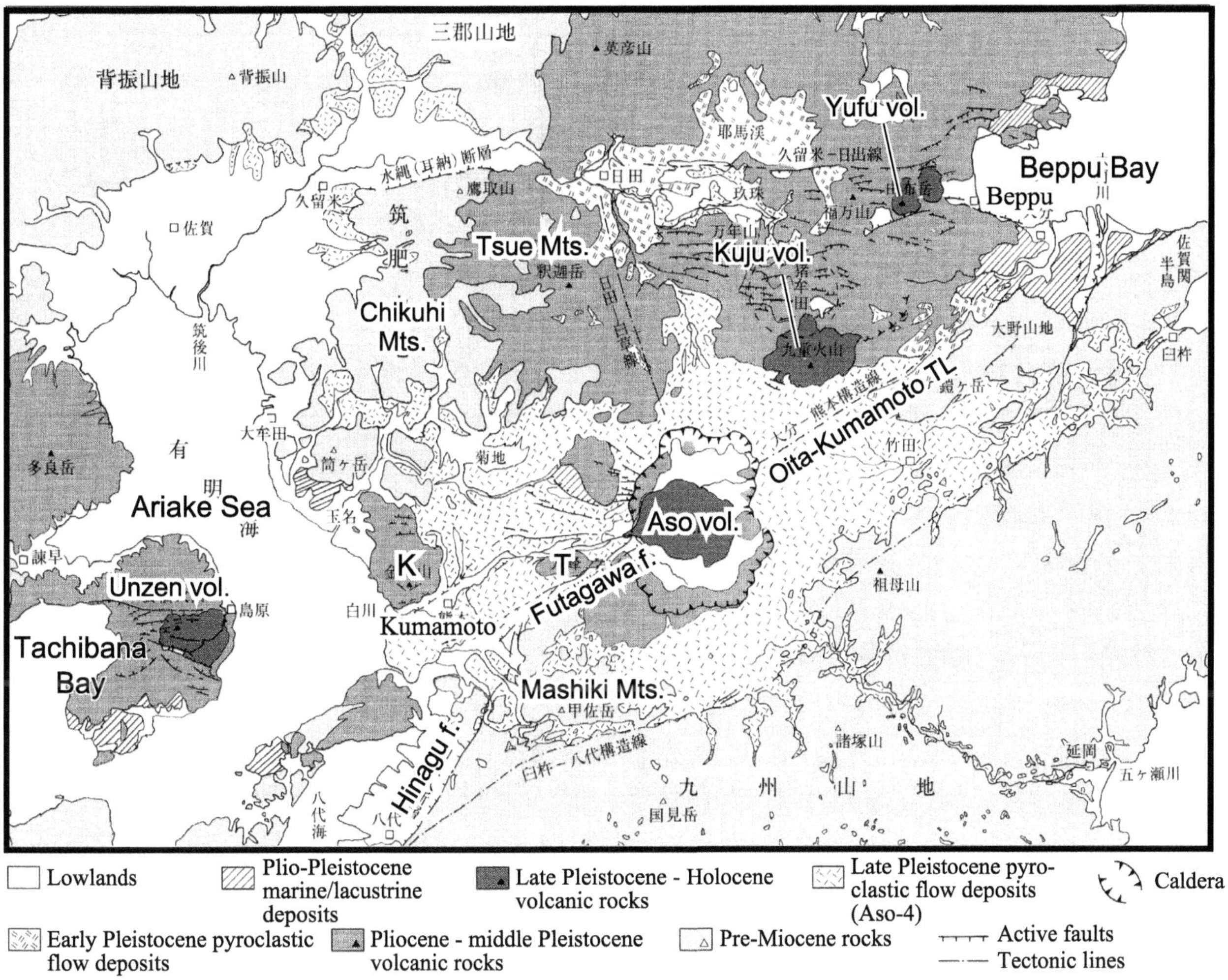

Fig. 1.2 Geology and geomorphology of central Kyushu modified from Machida et al. (2001). K: Kinpozan Mountain, T: Takayubaru Upland composed of lava flow deposits from the Ohmine Volcano

wide in the east–west direction. There are more than 18 volcanic cones in the caldera, such as the Nakadake and Kijimadake. Some of the volcanic cones have erupted in the Holocene. The rocks ejected from the Aso Volcano show a wide variation in chemical composition from basalt to rhyolite (SiO_2, 49–74%). Thick piles of pyroclastic flow deposits erupted from the Aso Volcano since ~270 ka have formed gently sloping hills around the caldera (Figs. 1.2 and 1.4). There are four significant pyroclastic flow deposits; Aso-1 (250–270 ka), Aso-2 (~140 ka), Aso-3 (120–140 ka), and Aso-4 (85–90 ka) (Machida and Arai 2003). Pyroclastic flow deposits of these massive eruptions are identified widely in central Kyushu, whereas the related air-fall ashes are distributed widely on the Japanese islands and surrounding areas, making them essential tephra layers of middle to late Pleistocene age. Of these, the Aso-4 pyroclastic flow deposits are responsible for most of the gently sloping hills around the volcano, including part of the Takuma Upland east of downtown Kumamoto (Fig. 1.4). The Ohmine Volcano west of the caldera is a monogenic volcano that erupted between Aso-3 and Aso-4 eruptions. The Takayubaru Upland north of the Futagawa fault comprises hornblende andesite lava erupted from the Ohmine Volcano (Fig. 1.2).

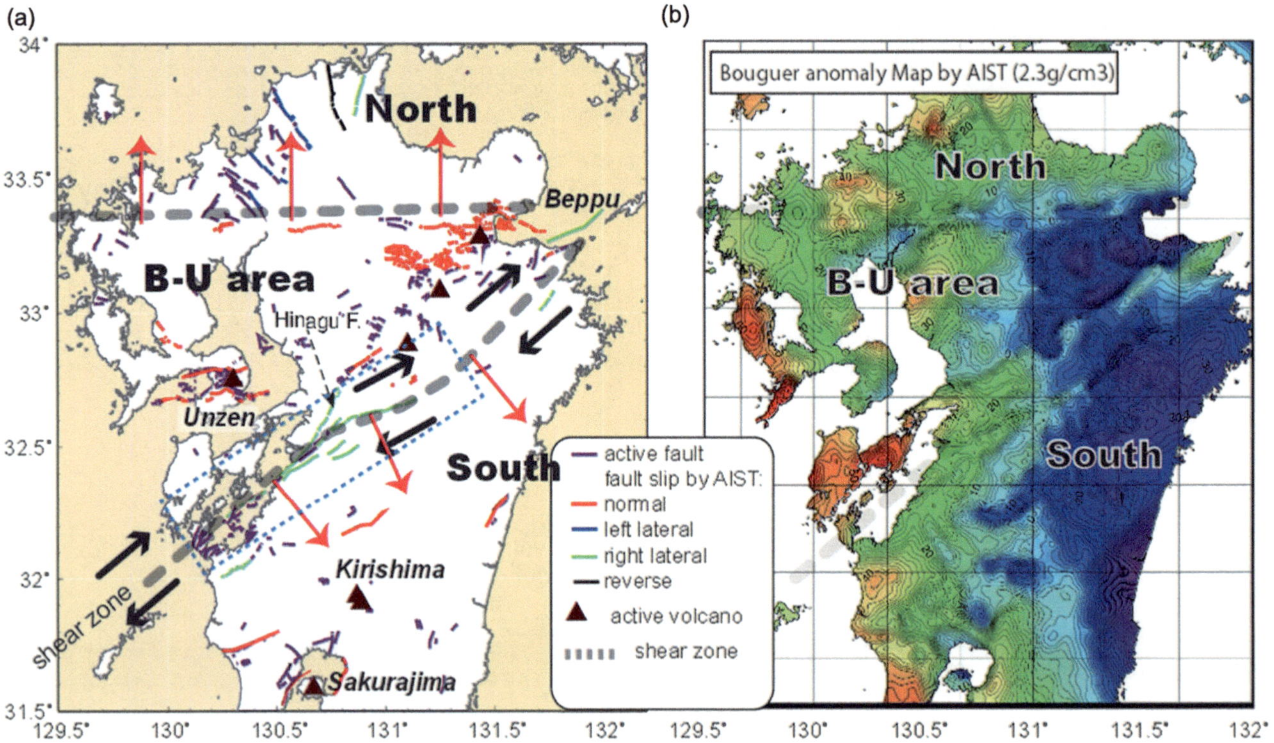

Fig. 1.3 **a** Schematic illustration of the stress field and tectonic divisions of Kyushu Island. Black arrows indicate the slip directions along the shear zone that bounds the BSGZ (B-U area) on the south. Red arrows show extensional force acting on the B-U area. **b** Bouguer gravity anomaly map assuming a terrain density of 2.3 g/cm^3. From Matsumoto et al. (2015)

The Kumamoto Plain is located at the lower reaches of the two major rivers: the Shirakawa River that drains from the Aso caldera basin and Midorikawa River that originates from the southern flank of the caldera and flows through the mountains to the southwest including the Mashiki Mountains. The Futagawa fault marks the southern margin of the plain (Figs. 1.2 and 1.4). Based on the depth to the Miyuki Formation, the marine oxygen-isotope stage (MIS) 5c or 5e marine sediments, a subsidence rate of 0.45–0.90 mm/yr was estimated for the Kumamoto Plain, with a higher subsidence rate to the west (Ishizaka et al., 1995). Despite this rapid subsidence, vast volcanoclastic deposits from the Aso Volcano have prohibited the sea from transgressing to the Kumamoto Plain.

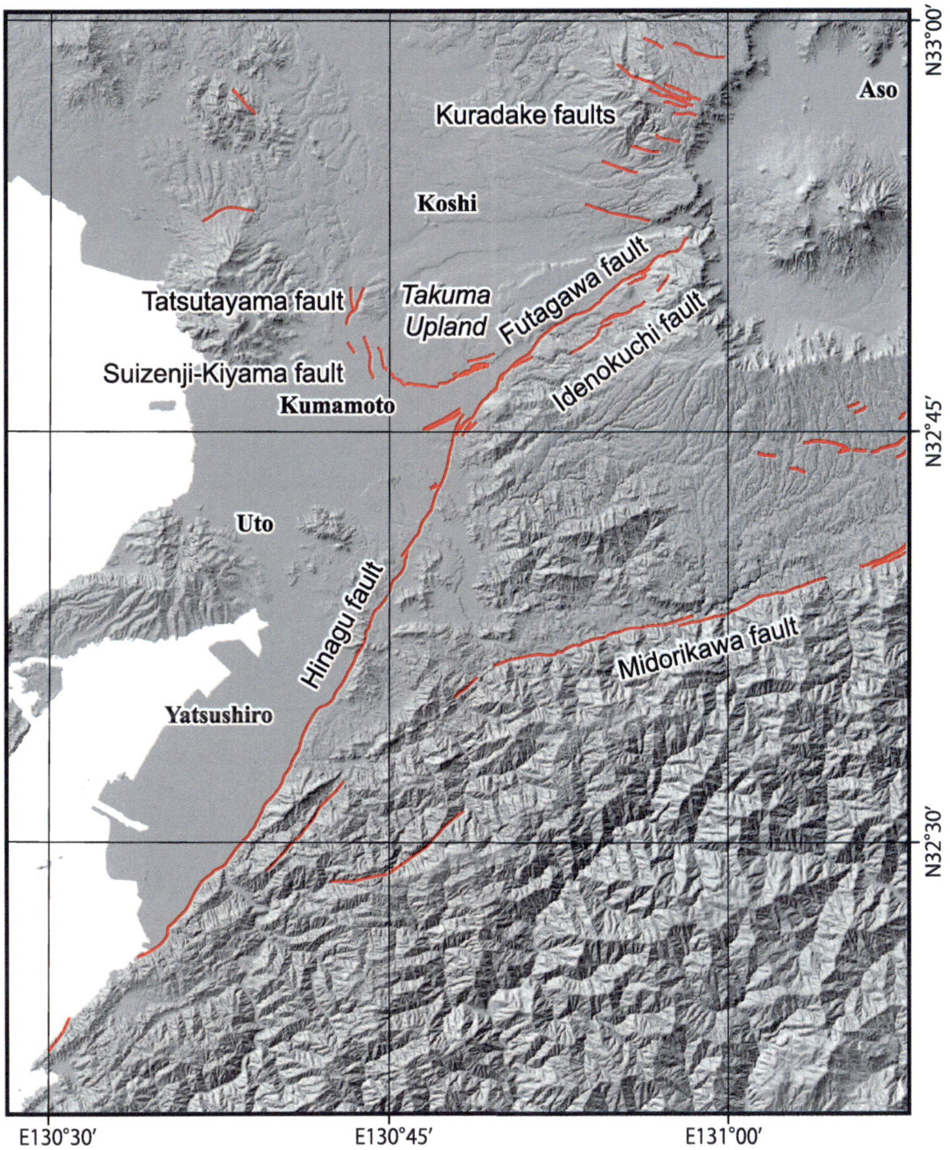

Fig. 1.4 Topography and active faults (red lines) in and around the epicentral area of the 2016 Kumamoto earthquake sequence. Active fault traces are from Imaizumi et al. (2018) and Kumahara et al. (2017)

References

Eguchi T, Uyeda S (1983) Seismotectonics of the Okinawa Trough and Ryukyu Arc. Memoir Geol Soc China 5:1–27

Geological Society of Japan (ed) (2010) Reginal geology of Japan-Kyushu and Okinawa. Asakura Shoten, Tokyo

Imaizumi T, Miyauchi T, Tsutsumi H, Nakata T (eds) (2018) Digital active fault map of Japan [revised edition]. University of Tokyo Press, Tokyo

Ishizaka S, Iwasaki Y, Hase Y, Watanabe K, Iwauchi A, Taziri M (1995) Subsidence rate and sediments of the last interglacial epoch in the Kumamoto Plain, Japan. Quarter Res (Daiyonki Kenkyu) 34:335–344

Kamata H (1989) Volcanic and structural history of the Hohi volcanic zone, central Kyushu, Japan. Bull Volcanol 51:315–332

Kumahara Y, Okada S, Kagohara K, Kaneda H, Goto H, Tsutsumi H (2017) Active fault map 'Kumamoto" [revised edition]. Geospatial Information Authority of Japan D1-No.868, Geospatial Information Authority of Japan, Tokyo

Machida H, Arai F (2003) Atlas of tephra in and around Japan [revised edition]. University of Tokyo Press, Tokyo

Machida H, Ota Y, Kawana T, Moriwaki H, Nagaoka S (eds) (2001) Regional geomorphology of the Japanese islands, vol 7 Geomorphology of Kyushu and the Ryukyus. University of Tokyo Press, Tokyo

Matsumoto Y (1979) Some problems on volcanic activities and depression structure in Kyushu, Japan. Memoirs Geol Soc Japan 16:127–139

Matsumoto S, Nakao S, Ohkura T, Miyazaki M, Shimizu H, Abe Y, Inoue H, Nakamoto M, Yoshikawa S, Yamashita Y (2015) Spatial heterogeneities in tectonic stress in Kyushu, Japan and their relation to a major shear zone. Earth Planets Space. https://doi.org/10.1186/s40623-015-0342-8

Tada T (1993) Crustal deformation in central Kyusyu, Japan and its tectonic implication –Rifting and spreading of the Beppu-Shimabara Graben. Memoirs Geol Soc Japan 41:1–12

Seismicity and Crustal Movement

2

Hiroyuki Tsutsumi, Heitaro Kaneda, and Yasuhiro Kumahara

Abstract

This chapter summarizes the background seismicity and crustal movement of Kyushu Island, especially the Beppu–Shimabara graben zone. The Kumamoto area is located in a zone of high seismic activity in central Kyushu. Seismic and geodetic data illustrate pronounced north–south extension of the Beppu–Shimabara graben zone, consistent with the dense networks of east-trending active normal faults.

Keywords

2016 Kumamoto earthquake • Beppu–Shimabara graben zone • Seismicity • Crustal movement • Normal fault • N–S extension

Seismicity in and around Kyushu Island exhibits a sizeable regional variation, as shown in Fig. 2.1 (Geological Society of Japan 2010). The offshore earthquakes in the Hyuga-nada Sea are mostly subduction related, whereas onshore earthquakes and offshore earthquakes in the western sea are crustal earthquakes within the overlying plate. The earthquakes cluster in west-central Kyushu, i.e., in the area from Beppu Bay through the Unzen Volcano to the Amakusa-nada Sea and in the area from the Aso caldera to Yatsushiro Sea. The former coincides with the Beppu-Shimabara graben zone (BSGZ), and the latter is related to the Futagawa-Hinagu fault zone. Focal mechanism solutions for the crustal earthquakes on Kyushu are predominantly strike-slip, normal, and their combinations. Figure 2.2 shows the directions of the P-axis and T-axis for the crustal earthquakes, indicating a stress field with E–W compression and N–S extension. The direction of the T-axis rotates counterclockwise from almost N–S in northern Kyushu to NW-SE in southern Kyushu. Matsumoto et al. (2016) demonstrated that the depth above which 95% of crustal events occur (D95 depth) in 1993–2013 is shallower than 10 km at the Beppu area in the eastern BSGZ. In contrast, seismic activity around the Kumamoto area extends to depths of 15–18 km.

Trilateration measurements conducted in 1891–1893 and 1989–1991 indicated >10 ppm N–S extension in ~100 years (>0.1 ppm/yr) for the BSGZ (Geological Society of Japan 2010). The recent GNSS observations suggested a bit smaller N–S extension of ~0.05 ppm/yr in 1996–2000 (Sagiya 2004). These observations are consistent with the general characteristics of the region's seismicity and dense distribution of east-trending normal faults in the BSGZ.

Supplementary Information
The online version contains supplementary material available at (https://doi.org/10.1007/978-981-19-1150-7_2).
Note
The supplementary material of this book contains geographical information system (GIS) and spreadsheet files of the surface ruptures associated with the mainshock of the 2016 Kumamoto earthquake sequence

H. Tsutsumi (✉)
Department of Environmental Systems Science,
Faculty of Science and Engineering, Doshisha University,
1-3, Tatara-Miyakodani, Kyotanabe, Kyoto, 610-0394, Japan
e-mail: htsutsum@mail.doshisha.ac.jp

H. Kaneda
Department of Civil and Environmental Engineering,
Chuo University, 1-13-27, Kasuga, Bunkyo-ku,
Tokyo, 112-8551, Japan
e-mail: kaneda.00c@g.chuo-u.ac.jp

Y. Kumahara
Graduate School of Humanities and Social Sciences,
Hiroshima University, 1-1-1, Kagamiyama, Higashi-Hiroshima,
Hiroshima, 739-8524, Japan
e-mail: kumakuma@hiroshima-u.ac.jp

Y. Kumahara et al. (eds.), *Surface Ruptures Associated with the 2016 Kumamoto Earthquake Sequence in Southwest Japan*,
Advances in Geological Science, https://doi.org/10.1007/978-981-19-1150-7_2

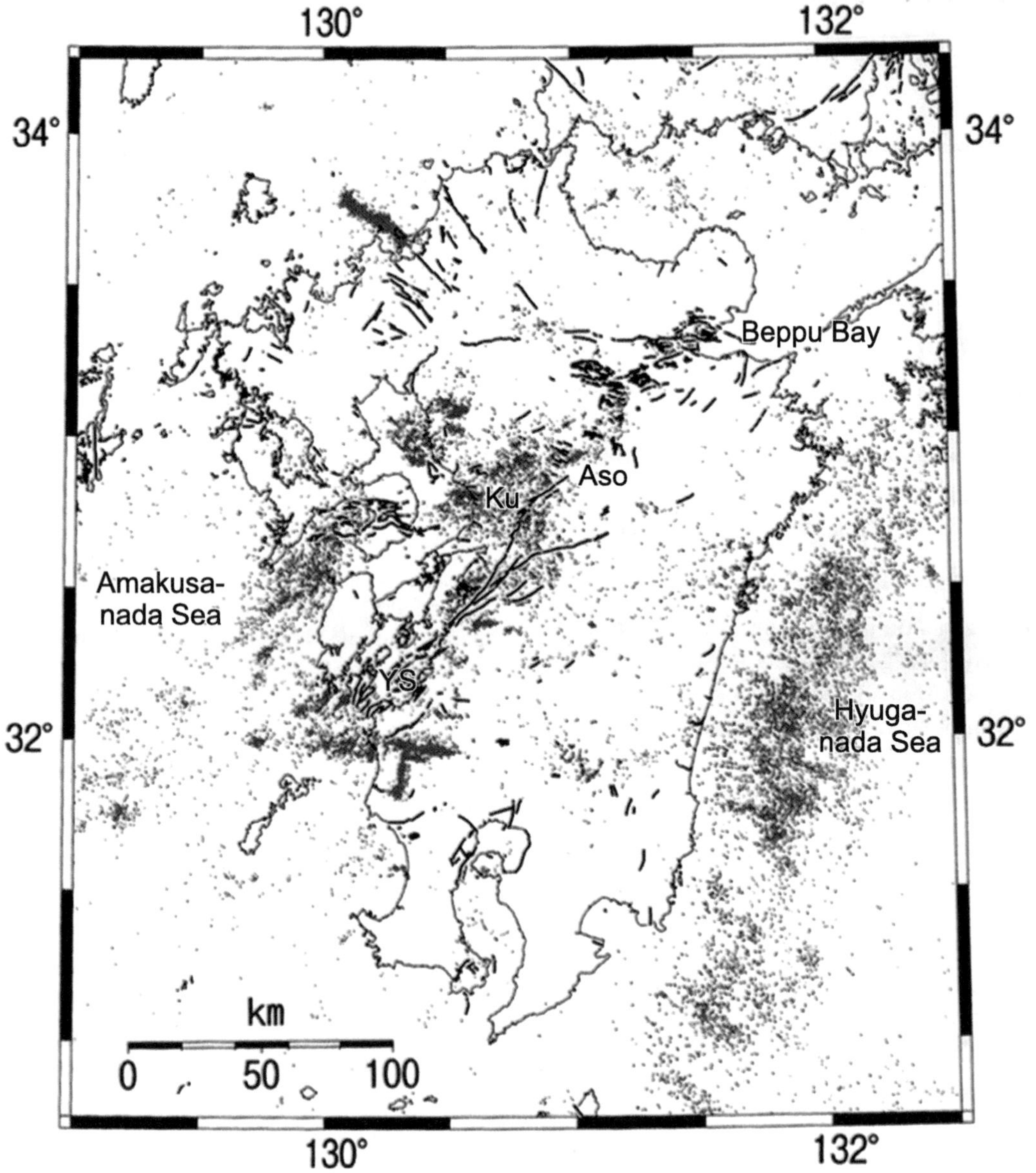

Fig. 2.1 Distribution of earthquakes (M $\geq$ 1) shallower than 30 km depth from 1996 through 2005. Solid lines denote active faults. Ku: Kumamoto, YS: Yatsushiro Sea. Modified from Geological Society of Japan (2010)

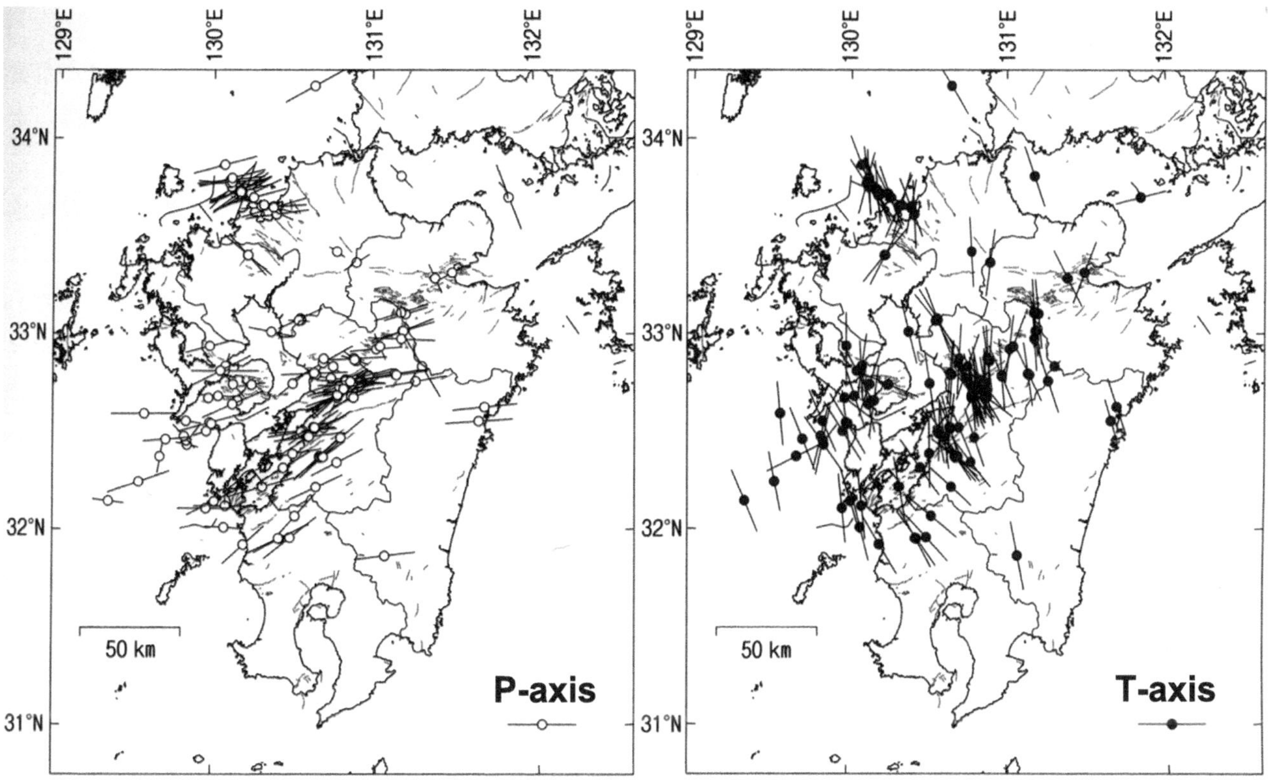

Fig. 2.2 Directions of P-axis (left) and T-axis (right) inferred from shallow crustal earthquakes occurred in Kyushu and surrounding areas. Modified from Geological Society of Japan (2010)

References

Geological Society of Japan (ed) (2010) Regional geology of Japan –Kyushu and Okinawa. Asakura Shoten, Tokyo

Matsumoto S, Nishimura T, Ohkura T (2016) Inelastic strain rate in the seismogenic layer of Kyushu Island, Japan. Earth Planets Space. https://doi.org/10.1186/s40623-016-0584-0

Sagiya T (2004) A decade of GEONET: 1994–2003 -the continuous GPS observation in Japan and its impact on earthquake studies-. Earth Planets Space. https://doi.org/10.1186/BF03353077

Active Faults and Paleoseismicity

3

Hiroyuki Tsutsumi, Heitaro Kaneda, Yasuhiro Kumahara,
and Yoshiya Iwasa

Abstract

This chapter briefly describes active faults distributed in the epicentral area of the Kumamoto earthquake sequence, many of which were reactivated in 2016. We summarize the location, length, and slip sense of these faults. We also introduce paleoseismic data for these active faults obtained before and after the Kumamoto earthquake sequence. It is now known that the recurrence interval of the surface-rupturing earthquakes on the Futagawa fault is ~2000 years, considerably shorter than the estimate before the 2016 earthquake.

Supplementary Information
The online version contains supplementary material available at (https://doi.org/10.1007/978-981-19-1150-7_3).
Note
The supplementary material of this book contains geographical information system (GIS) and spreadsheet files of the surface ruptures associated with the mainshock of the 2016 Kumamoto earthquake sequence

H. Tsutsumi (✉)
Department of Environmental Systems Science, Faculty of Science and Engineering, Doshisha University, 1-3, Tatara-Miyakodani, Kyotanabe, Kyoto, 610-0394, Japan
e-mail: htsutsum@mail.doshisha.ac.jp

H. Kaneda
Department of Civil and Environmental Engineering, Chuo University, 1-13-27, Kasuga, Bunkyo-ku, Tokyo, 112-8551, Japan
e-mail: kaneda.00c@g.chuo-u.ac.jp

Y. Kumahara · Y. Iwasa
Graduate School of Humanities and Social Sciences, Hiroshima University, 1-1-1, Kagamiyama, Higashi-Hiroshima, Hiroshima, 739-8524, Japan
e-mail: kumakuma@hiroshima-u.ac.jp

Y. Iwasa
e-mail: yiwasa067@hiroshima-u.ac.jp

Keywords

2016 Kumamoto earthquake • Active fault • Futagawa–Hinagu fault zone • Paleoseismology • Recurrence interval

A clustering of east-striking normal faults around volcanos characterizes the Beppu–Shimabara graben zone. Active faults are especially densely clustered in the Beppu–Kuju area, the Aso-Kumamoto area, and the Unzen area (Fig. 1.1). Many of the faults are short normal faults that run subparallel to one another and form grabens. Figure 1.4 in Chap. 1 shows detailed active fault traces in the epicentral area of the Kumamoto earthquake based on Imaizumi et al. (2018) and Kumahara et al. (2017a). The major faults in the area are (1) NE-trending Futagawa and Idenokuchi faults, (2) NNE-trending Hinagu fault, (3) WNW-trending Kuradake faults, (4) NNW–ENE-trending Suizenji–Kiyama fault, and (5) NE-trending Tatsutayama fault.

Along the southern margin of the Kumamoto Plain is the right-lateral Futagawa fault with a north-side-down component of vertical displacement. The fault extends from the western margin of the Aso caldera to the southeast of downtown Kumamoto for ~19 km. Headquarters for Earthquake Research Promotion (2013) identified the 47-km-long possible western extension of the Futagawa fault through north of Uto into the Ariake Sea based on subsurface geologic and gravity data (not shown in Fig. 1.4). The Idenokuchi fault runs subparallel to the Futagawa fault for ~10 km with a predominant northwest-side-down displacement component.

The right-lateral Hinagu fault is about 81 km long, including submarine faults in the Yatsushiro Sea and marks a topographic boundary between the Yatsushiro Plain and Kyushu Mountains. The Hinagu fault is divided into three

Y. Kumahara et al. (eds.), *Surface Ruptures Associated with the 2016 Kumamoto Earthquake Sequence in Southwest Japan*,
Advances in Geological Science, https://doi.org/10.1007/978-981-19-1150-7_3

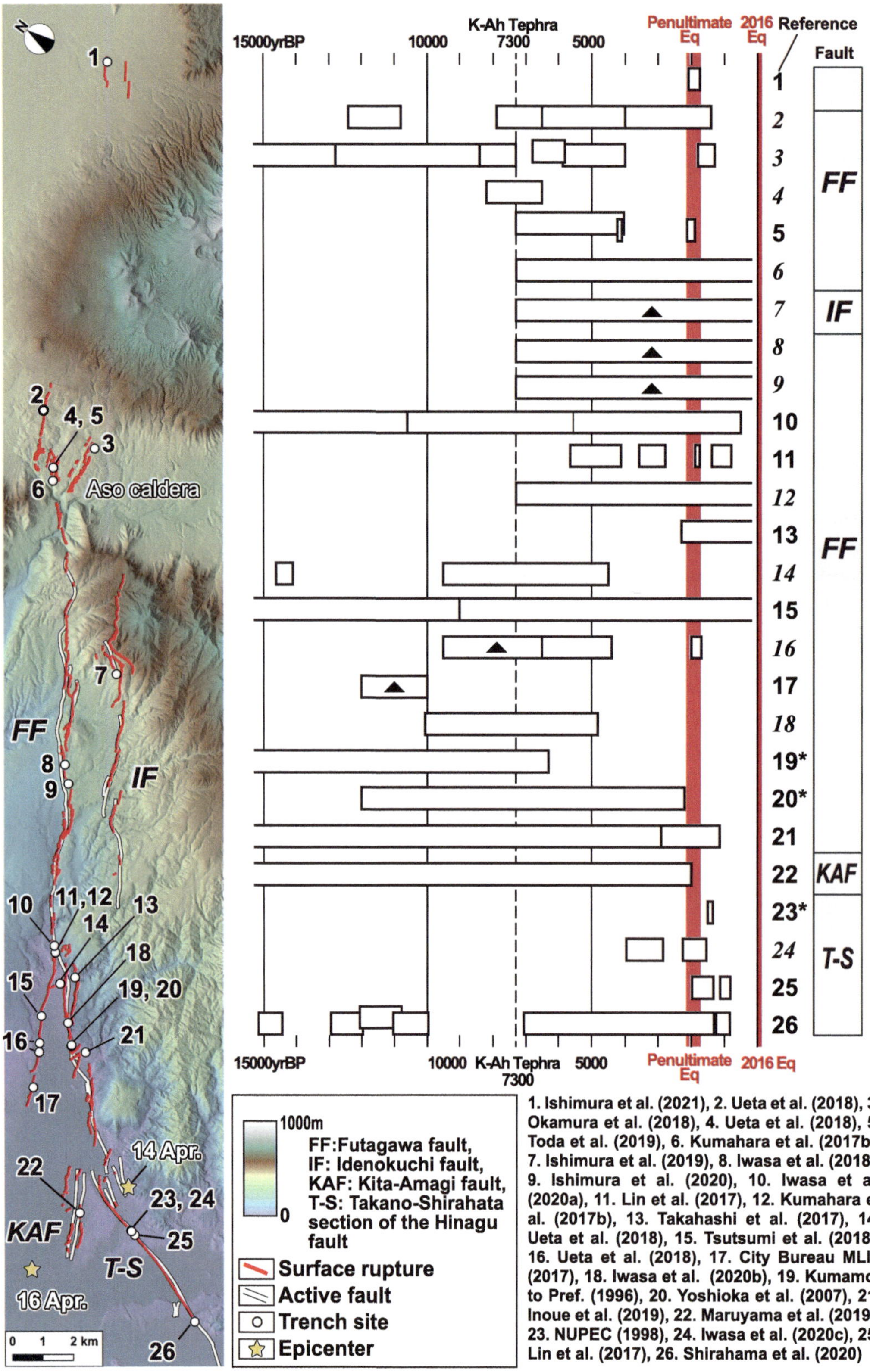

Fig. 3.1 Spatiotemporal diagram of the past faulting history on the surface ruptures of the 2016 Kumamoto earthquake, as revealed by trench excavation surveys as of June 2021. The black triangle indicates that the event occurred twice during the time interval shown by the box. Reference numbers with bold letters are studies published as papers or reports, those with italic letters are studies published as meeting or conference abstracts, and those with asterisks are studies published before the 2016 Kumamoto earthquake sequence. The K-Ah tephra erupted ~7300 years ago from a submarine caldera south of Kyushu is an excellent time marker commonly found in Kyushu area and elsewhere in southwest-central Japanese Archipelago

segments (sections) based on geometry and paleoseismic data: the Takano-Shirahata, Hinagu, and Yatsushiro Sea sections (Headquarters for Earthquake Research Promotion 2013). The Futagawa and Hinagu faults are often interpreted as a single continuous fault system referred to as the Futagawa–Hinagu fault zone (e.g., Nakata and Imaizumi 2002; Imaizumi et al. 2018) because the fault traces smoothly connect with no distinct gap, although the overall strike changes as much as ~50° along the two faults.

The Kuradake faults are a series of WNW-trending short (less than 6 km long) subparallel normal faults cutting the pyroclastic-flow surfaces of Aso-1–4 on the western flank of the Aso caldera. Although the surfaces are older than 85–90 ka, the heights of the fault scarps are less than 30 m.

The Suizenji–Kiyama fault is a ~11-km-long active fault that has a curved trace convex to the southwest. The western portion of the fault offsets fluvial terraces down to the west and was identified after the 2016 earthquake sequence (Goto et al. 2017). The eastern part, first identified by Watanabe et al. (1979), bounds the southern margin of the Takuma upland and forms a graben structure called the Kiyama–Kashima graben (Watanabe et al. 1979) with the Futagawa fault bounding its southern margin.

The Tatsutayama fault is a short (~3 km long) active fault at the western foot of the Tatsutayama Mountain. The fault may be the source fault of the 1889 M 6.3 earthquake (Kubodera et al. 1988), the last devastating earthquake before the 2016 event that struck the Kumamoto area, because the damage was particularly concentrated along the fault trace. The 1889 earthquake reportedly claimed 19 lives (Usami et al. 2013).

Several paleoseismic trenching studies had been already conducted on the Futagawa and Hinagu faults before the 2016 earthquake sequence and far more were carried out since 2016. Before the earthquake sequence, past seismic events of these faults were poorly constrained. The timing of the last surface-rupturing event on the Futagawa fault was only constrained as sometime during 6.9–2.2 ka (Headquarters for Earthquake Research Promotion 2013). The average recurrence interval was also loosely constrained as 8100–2600 years. The paleoseismic activity of the Hinagu fault was better constrained. The timing of the last surface-rupturing events on the three sections of the fault was constrained as sometime between 1.6 and 1.2 ka (northern part: Takano-Shirahata section), 8.4–2.0 ka (central part: Hinagu section), and 1.7–0.9 ka (southern part: Yatsushiro Sea section) (Headquarters for Earthquake Research Promotion 2013).

After the 2016 earthquake sequence, trench excavation surveys were conducted at 23 locations as of June 2021 along the surface ruptures (Fig. 3.1). In all surveys, displacement and deformation beneath the surface rupture were larger and more significant than those induced by the 2016 main shock, indicating repeated fault movement. Most of the trenching studies dated the penultimate event at ~2000 years ago, suggesting that rupture extent during the penultimate event was similar to that of the 2016 event. It is not clear, however, if the extent of the earlier events were also similar to those of the last two events because the timings of those events are only loosely constrained and hard to be correlated. Nonetheless, Toda et al. (2019) conducted a trench survey on the northern part of Futagawa fault inside the Aso caldera and proposed a narrowly defined age of 4237 cal BP–4.1 cal ka BP for the event before the penultimate event. Ishimura (2019) estimated both the vertical and right-lateral slip rates of the Futagawa fault at more than 1 mm/yr. If a single earthquake generates a right-lateral offset of ~2 m as in the 2016 main shock and slip rate is 1 mm/yr, the recurrence interval of faulting is calculated to be ~2000 years, which is consistent with the event-timing estimate of Toda et al. (2019). In any case, the extensive trench excavations after the 2016 earthquake sequence clarified that the rupture interval of the Futagawa fault is well shorter than previously evaluated (8100–2600 years; Headquarters for Earthquake Research Promotion 2013).

References

City Bureau, Ministry of Land, Infrastructure, Transport and Tourism (2017) Final report on safety measures for urban reconstruction in Mashiki Town after the Kumamoto earthquake. Ministry of Land, Infrastructure, Transport and Tourism, Tokyo

Goto H, Tsutsumi H, Toda S, Kumahara Y (2017) Geomorphic features of surface ruptures associated with the 2016 Kumamoto earthquake in and around the downtown of Kumamoto City, and implications on triggered slip along active faults. Earth Planets Space 69:26. https://doi.org/10.1186/s40623-017-0603-9

Headquarters for Earthquake Research Promotion (2013) Evaluation of the Futagawa fault zone and Hinagu fault zone. https://www.jishin.go.jp/main/chousa/katsudansou_pdf/93_futagawa_hinagu_2.pdf. Accessed 20 Aug 2021

Imaizumi T, Miyauchi T, Tsutsumi H, Nakata T (eds) (2018) Digital active fault map of Japan [revised edition]. University of Tokyo Press, Tokyo

Inoue N, Kitada N, Shibuya N, Omata M, Takahama T, Tonagi M, Irikura K (2019) Probabilistic evaluation of off-fault displacements of the 2016 Kumamoto earthquake. Pure Appl Geophys. https://doi.org/10.1007/s00024-019-02345-7

Ishimura D (2019) Co-seismic vertical displacement associated with the 2016 Kumamoto earthquake (Mw7.0) and activity of the Futagawa fault around Futa, Nishihara Village, Kumamoto Prefecture. Active Fault Res 50:33–34

Ishimura D, Tsutsumi H, Takahashi N, Oda R, Matsukaze J, Kaneda H, Kobayashi M, Kumahara Y, Kobayashi M, Ichihara T (2019) Paleo-seismological survey on the Idenokuchi fault, Nishihara Village, Kumamoto Prefecture. In: Abstracts of the Japan Geoscience Union Meeting 2019, Makuhari Messe, Chiba, 26–30 May 2019

Ishimura D, Iwasa Y, Takahashi N, Tadokoro R, Oda R, Kajii K, Matsukaze J, Ishizawa T, Tsutsumi H (2020) Paleo-seismological survey on the Futagawa fault at Futa, Nishihara Village, Kumamoto

Prefecture. In: Abstracts of the Japan Geoscience Union, American Geophysical Union Joint Meeting 2020, Makuhari Messe, Chiba, 24–28 May 2020

Ishimura D, Tsutsumi H, Toda S, Fukushima Y, Kumahara Y, Takahashi N, Ichihara T, Takada K (2021) Repeated triggered ruptures on a distributed secondary fault system: an example from the 2016 Kumamoto earthquake, southwest Japan. Earth Planets Space 73:39. https://doi.org/10.1186/s40623-021-01371-x

Iwasa Y, Kumahara Y, Goto H, Torii M, Ishimura D (2018) Faulting history of the northeastern part of Futagawa-Hinagu fault zone based on trench survey at Komori, Nishihara Village, Kumamoto. In: Abstract of the Japanese Society for Active Fault Studies Fall Meeting, Hiroshima University, Hiroshima, 24–26 Nov 2017

Iwasa Y, Kumahara Y, Goto H, Nakata T (2020a) Detailed mapping of surface ruptures associated with the 2016 Kumamoto earthquake and faulting history of the conjugated fault in Dozon, Mashiki Town, Kumamoto Prefecture. Active Fault Res 52:1–8

Iwasa Y, Kumahara Y, Goto H, Torii M (2020b) Paleoseismic history and horizontal slip of the Futagawa–Hinagu fault zone based on paleoseismological and tectonic geomorphological survey at Hirata, Mashiki Town, Kumamoto Prefecture. In: Abstract of the Japan Geoscience Union, American Geophysical Union Joint Meeting 2020, Makuhari Messe, Chiba, 24–28 May 2020

Iwasa Y, Kumahara Y, Goto H, Hosoya T, Takeuchi S, Sato T, Sumitani Y, Nishiguchi S (2020c) Faulting history of the 2016 surface rupture in the boundary area of the Futagawa and Hinagu fault zones at Takagi, Mifune Town, Kumamoto Prefecture. In: Abstract of the Japanese Society for Active Fault Studies Fall Meeting, Online, 30–31 Nov 2020

Kubodera A, Omote S, Yokoyama S, Watanabe K, Miyazaki M, Narahashi H (1988) Reevaluation of the 1889 Kumamoto earthquake. West Reg Div Rep Jpn Group Study Nat Disaster 5:1–6

Kumahara Y, Okada S, Kagohara K, Kaneda H, Goto H, Tsutsumi H (2017a) Active fault map 'Kumamoto" [revised edition]. Geospatial Information Authority of Japan D1-No.868, Geospatial Information Authority of Japan, Tokyo

Kumahara Y, Torii M, Nakata T, Goto H, Iwasa Y, Suzuki Y, Watanabe M, Yoda S, Takahashi N, Okuno M (2017b) Fault history of the northeastern part of Futagawa-Hinagu fault zone based on trench survey at Dozon, Mashiki Town and at Kawayo, Minami-Aso Village (preliminary result). In: Abstract of the Japanese Society for Active Fault Studies Fall Meeting, Hiroshima University, Hiroshima, 24–26 Nov 2017

Kumamoto Prefecture (1996) Research report on the Futagawa fault and the Tattayama fault. Kumamoto Prefecture, Kumamoto

Lin A, Chen P, Satsukawa T, Sado K, Takahashi N, Hirata S (2017) Millennium recurrence interval of morphogenic earthquakes on the seismogenic fault zone that triggered the 2016 Mw 7.1 Kumamoto earthquake, Southwest Japan. Bull Seismol Soc Am 107:2687–2702. https://doi.org/10.1785/0120170149

Maruyama T, Saito M, Komine Y, Kamedaka M (2019) Paleoseismological trenching and detailed mapping of the surface rupture appeared along the Kita-amagi fault during the 2016 Kumamoto earthquake at Shimada, Mashiki Town, Kumamoto Prefecture, southwest Japan. Active Fault Res 50:13–31

Nakata T, Imaizumi T (eds) (2002) Digital active fault map of Japan. University of Tokyo Press, Tokyo

Nuclear Power Engineering Corporation (1998) Report on the feasibility study on the location of nuclear power plants in 1997, vol 1. Nuclear Power Engineering Corporation, Tokyo

Okamura Y, Miyashita Y, Abe S, Awata Y, Azuma T, Togo T, Shirahama Y, Maruyama T, Ogami T, Imura R, Tsutsumi H, Goto H, Kumahara Y, Torii M (2018) Survey of detailed position and shape of active faults to understand the fault segments and observation to reveal the paleoseismic history and slip rates. In: Research report of a comprehensive active fault survey after the 2016 Kumamoto earthquake, 2017 fiscal year. Ministry of Education, Culture, Sports, Science and Technology and Kyushu University. https://www.jishin.go.jp/main/chousakenkyuu/kumamoto_sogochousa/h29/h29kumamoto_sogochousa_3_1.pdf. Accessed 6 May 2021

Shirahama Y, Miyashita Y, Kametaka M, Suzuki Y, Miyairi Y, Yokoyama Y (2020) Detailed paleoseismic history of the Hinagu fault zone revealed by the high-density radiocarbon dating and trenching survey across a surface rupture of the 2016 Kumamoto earthquake, Kyushu, Japan. Island Arc. https://doi.org/10.1111/iar.12376

Takahashi N, Ishimura D, Toda S, Nakata T, Watanabe M (2017) Vertical slip rate on a normal fault co-ruptured with the Futagawa fault at the 2016 Kumamoto earthquake: a fault outcrop at Shimojin, Mashiki Town in Kumamoto Prefecture. Active Fault Res 46:27–32

Toda S, Torii M, Okuno M, Konno A, Ono H, Takahashi N (2019) Evidence for Holocene paleoseismic events on the 2016 Kumamoto earthquake rupture zone within the Aso caldera: a trench excavation survey at Kurokawa, the town of Minami-Aso, southwest Japan. Active Fault Res 51:13–25

Tsutsumi H, Toda S, Goto H, Kumahara K, Ishimura D, Takahashi N, Taniguchi K, Omata M, Kohriya Y, Gomi M, Asano K, Iwata T (2018) Paleoseismic trenching across the surface rupture of the 2016 Kumamoto earthquake at Jichu, Mashiki Town, Kumamoto Prefecture. Active Fault Res 49:31–39

Ueta K, Miyawaki R, Iemura K, Yokoyama T, Miyawaki A (2018) Paleoseismological study on surface fault ruptures produced by the 2016 Kumamoto earthquake. In: Abstracts of Japan Geoscience Union Meeting 2018, Makuhari Messe, Chiba, 20–24 May 2018

Usami T, Ishii H, Imamura T, Takemura M, Matsu'ura RS (2013) Materials for comprehensive list of destructive earthquakes in Japan, 599–2012 [revised in 2013]. University of Tokyo Press, Tokyo

Watanabe K, Momikura Y, Tsuruta K (1979) Active faults and parasitic eruption centers on the west flank of Aso caldera, Japan. Quat Res (Daiyonki Kenkyu) 18:89–101

Yoshioka T, Shintani K, Iemura K, Miyawaki R (2007) Paleoseismicity of the Futagawa-Hinagu fault zone, central Kyushu, Japan. Annu Report Active Fault Paleoearthquake Res 7:241–258

Seismological and Geodetic Observations of the 2016 Kumamoto Earthquake Sequence

4

Hiroyuki Tsutsumi and Shinji Toda

Abstract

This chapter briefly introduces seismic and geodetic observations of the 2016 Kumamoto earthquake sequence to better understand the tectonic background of the coseismic surface ruptures. We describe focal mechanisms of the largest foreshock and mainshock, aftershock distributions, and source fault models based on various geophysical data. The severe seismic shaking of JMA intensity scale 7 was recorded twice at Mashiki Town, and its seismological interpretation is also provided.

Keywords

2016 Kumamoto earthquake • Foreshock • Mainshock • Source fault model • JMA seismic intensity scale • Strong ground motion

Supplementary Information
The online version contains supplementary material available at (https://doi.org/10.1007/978-981-19-1150-7_4).
Note
The supplementary material of this book contains geographical information system (GIS) and spreadsheet files of the surface ruptures associated with the mainshock of the 2016 Kumamoto earthquake sequence

H. Tsutsumi (✉)
Department of Environmental Systems Science, Faculty of Science and Engineering, Doshisha University, 1-3, Tatara-Miyakodani, Kyotanabe, Kyoto, 610-0394, Japan
e-mail: htsutsum@mail.doshisha.ac.jp

S. Toda
International Research Institute of Disaster Science (IRIDeS), Tohoku University, 468-1, Aoba, Aramaki, Aoba-ku, Sendai, Miyagi, 980-0845, Japan
e-mail: toda@irides.tohoku.ac.jp

The 2016 Kumamoto earthquake sequence started with the Mj = 6.5 (Mw = 6.2) earthquake on April 14 and its striking aftershocks. About 28 h later, the Mj = 7.3 (Mw = 7.0) earthquake occurred on April 16 with numerous widespread aftershocks (Fig. 4.1). Dense seismic networks deployed by the Japan Meteorological Agency (JMA) and the National Research Institute for Earth Science and Disaster Resilience (NIED) continuously monitored the space–time development of the seismic sequence (Fig. 4.1). Thanks to the rapid data acquisition system of satellite geodesy, detailed coseismic deformation associated with these two large events was also detected using paired data of the pre- and post-Kumamoto earthquake sequence. Numerous surface ruptures with a few centimeters of displacement were uncovered as interferogram fringe offsets. Here, we briefly summarize the seismological and geodetic observations of the earthquake sequence to provide the geophysical background of the surface ruptures described in the following chapters.

The Mj = 6.5 earthquake nucleated at a depth of ~12 km at 21:26:34 JST, April 14, 2016 (e.g., Kato et al. 2016). The JMA later designated this earthquake as the foreshock of the larger Mj = 7.3 mainshock on April 16. The aftershocks immediately following the foreshock occurred along the northernmost Hinagu fault (Fig. 4.1a). The CMT solution of the foreshock illustrates a right-lateral strike-slip mechanism that is consistent with the sense of motion of the Hinagu fault based on tectonic geomorphology. The GNSS and InSAR data document that a right-lateral slip occurred on a steeply WNW-dipping fault plane (Hinagu fault) during the foreshock (Kobayashi 2017; Kobayashi et al. 2018), whereas the relocation of the aftershocks (Shimizu et al. 2017) demonstrates that the source fault of the Mj = 6.5 earthquake steeply dips to the southeast. The largest aftershock (Mj = 6.4 (Mw = 6.1) on April 15) of the April 14 earthquake occurred on the WNW-dipping fault that appears to correspond to the Hinagu fault (Shimizu et al. 2017). Kato et al. (2016) inverted the amount of coseismic slip and fault

Y. Kumahara et al. (eds.), *Surface Ruptures Associated with the 2016 Kumamoto Earthquake Sequence in Southwest Japan*, Advances in Geological Science, https://doi.org/10.1007/978-981-19-1150-7_4

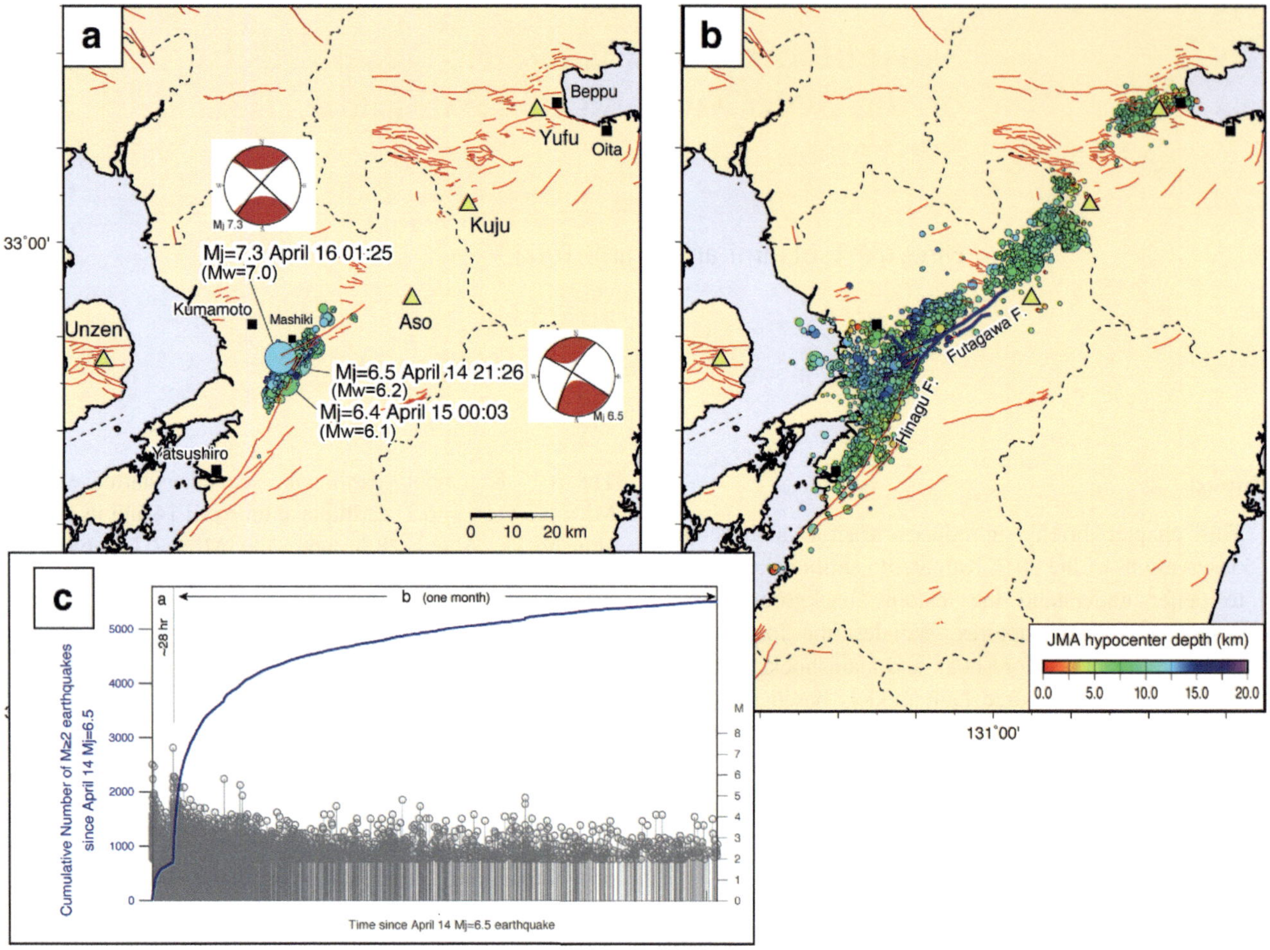

Fig. 4.1 Seismicity during the 2016 Kumamoto earthquake sequence. **a** The JMA CMT solutions for the Mj = 6.5 (Mw = 6.2) on April 14 and Mj = 7.3 (Mw = 7.0) on April 16, 2016 earthquakes, and aftershocks occurred during ~28 h between the two events. Red lines denote active faults. **b** One-month seismicity after the Mj = 7.3 earthquake. Blue lines are coseismic surface ruptures of the Mj = 7.3 event. **c** Seismic time series of the Kumamoto earthquake sequence. The blue curve indicates the cumulative number of M $\geq$ 2 earthquakes since the Mj = 6.5 earthquake on April 14, whereas each stem is an individual earthquake with height corresponding to its magnitude

length of the Mj = 6.5 foreshock as 1.1 m and 12.8 km, respectively. Small-scale surface ruptures appeared near the junction of the Futagawa and Hinagu faults in association with the foreshock and grew during the mainshock (Sugito et al. 2016).

About 28 h after the April 14 Mj = 6.5 event, the Mj = 7.3 mainshock occurred at 01:25:05 JST, April 16, at ~5 km west of the epicenter of the April 14 earthquake. The CMT solution of the mainshock shows a right-lateral strike-slip mechanism (Fig. 4.1a). From strong-motion waveform inversion, the rupture propagated mainly to the northeast and reached northwest of the Aso caldera with the maximum slip of ~4 m in a region 10–30 km northeast of the hypocenter (Fig. 4.2, Kubo et al. 2016). The other seismic inversion models (e.g., Asano and Iwata 2016) and aftershock distribution suggest that the source fault of the Mj = 7.3 earthquake dips to the northwest. The inverted seismic source models all display mixture of normal slip and right-lateral slip at depth in the eastern half of the Futagawa rupture zone (Fig. 4.2), which is also revealed by detailed geodetic inversion from InSAR analysis (Fig. 4.3; Geospatial Information Authority of Japan 2016; Kobayashi et al. 2018). On the surface, such an oblique motion on an NW-dipping source fault was reflected either remarkable vertical displacement component along the Futagawa fault (Kobayashi et al. 2018) or partitioned strike-slip on the Futagawa fault and normal slip along the Idenokuchi fault (Toda et al. 2016).

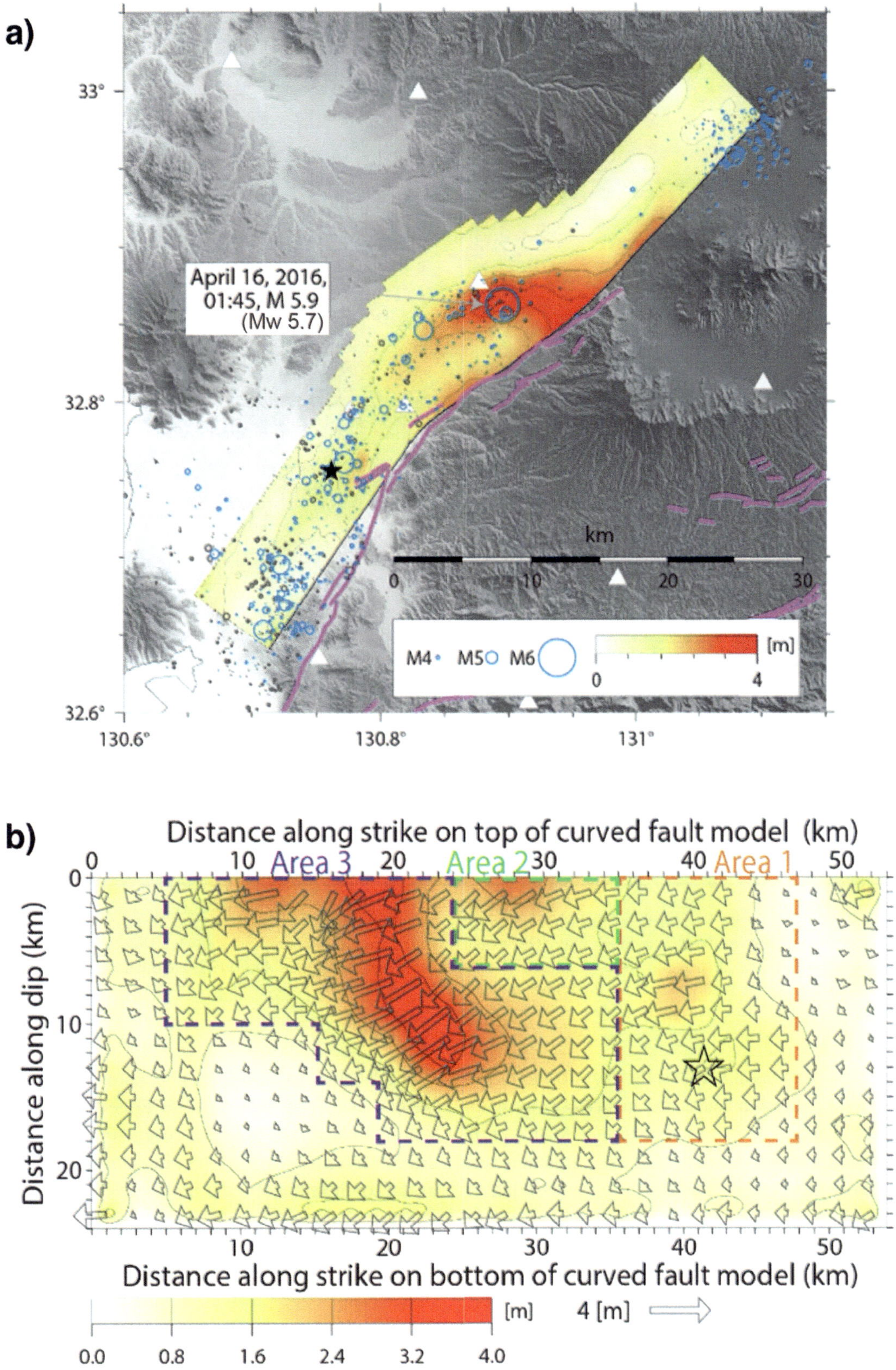

Fig. 4.2 Variable slip model of the Mj = 7.3 (Mw = 7.0) earthquake (modified after Kubo et al., 2016). **a** Map projection of the source fault and the total slip distribution. The slip contour interval is 0.8 m. The star indicates the hypocenter. Light blue and gray circles denote the M $\geq$ 1 aftershocks one day and one month after the mainshock, respectively. **b** Planar projection of the total slip distribution with vectors showing direction and amount of slip of the hanging wall side

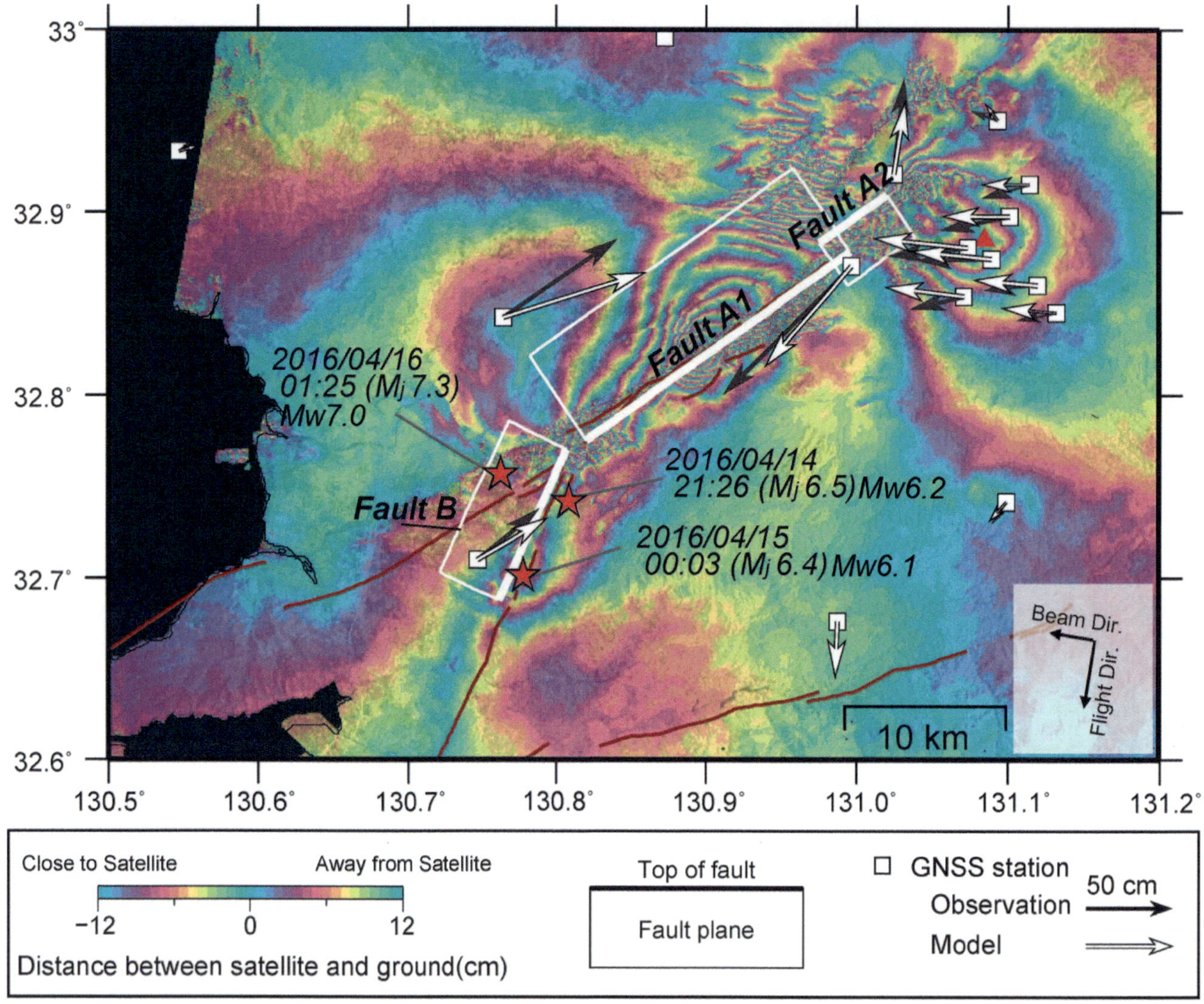

Fig. 4.3 Source fault model for the foreshock and mainshock, derived from GNSS data and InSAR analysis (modified after Geospatial Information Authority of Japan, 2016). The three rectangles denote surface projections of the modeled fault planes; the thick lines indicate the top of each fault plane. Black and white arrows at each GNSS station indicate horizontal surface displacements observed and predicted from the model, respectively. The background image is surface deformation from the ALOS-2 L-band data between 7 March 2016 and 18 April 2016

In addition to the seismicity along the northernmost Hinagu fault and Futagawa fault, the mainshock immediately triggered widespread aftershocks further to the north flank of the Aso volcano on the border of the Kumamoto and Oita Prefectures, including several M $\geq$ 5 earthquakes (Fig. 4.1b). One of the remarkable aftershocks was the Mj = 5.7 earthquake that occurred ~30 s later and ~80 km away from the mainshock epicenter, which brought strong shaking of JMA seismic intensity scale 6 lower to Beppu and Yufu areas in Oita Prefecture. This earthquake was interpreted as a remotely triggered earthquake (Yoshida 2016; Miyazawa 2016). The aftershocks were also extended to the southwest along the unruptured section of the Hinagu fault, sustaining high seismicity with low b-value (e.g., Nanjo et al. 2019), possibly associated with after-slip and viscoelastic deformation (Pollitz et al. 2017). As seismic

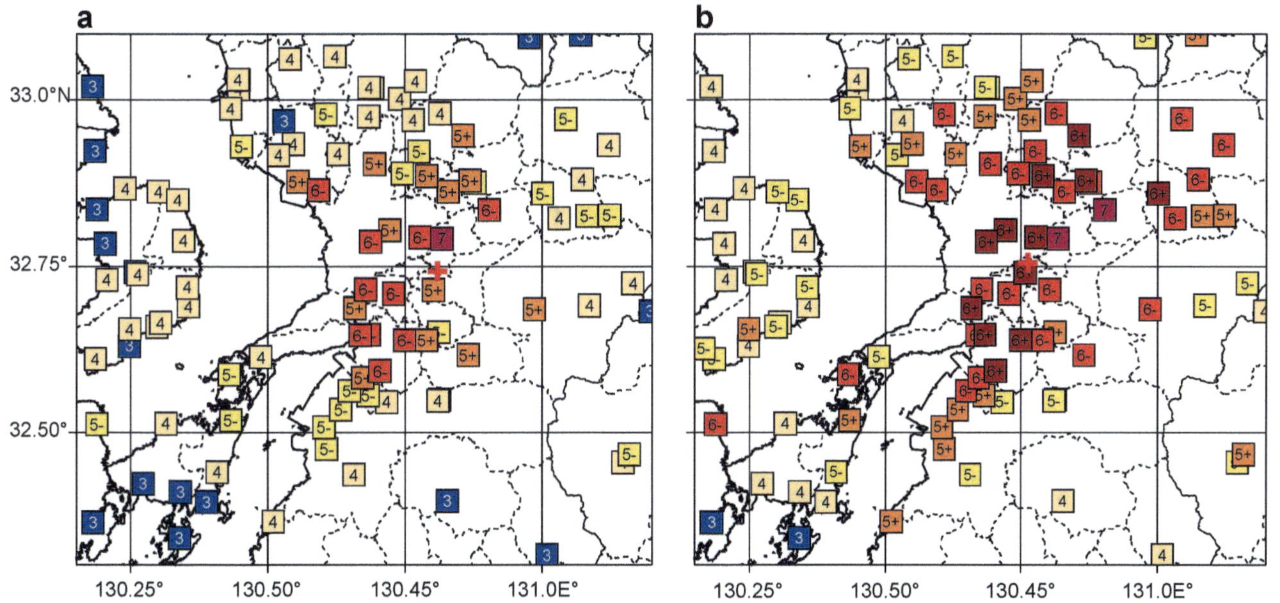

Fig. 4.4 Distributions of the JMA seismic intensity for the Mj = 6.5 (Mw = 6.2) foreshock (**a**) and Mj = 7.3 (Mw = 7.0) mainshock (**b**). From Kato et al. (2016)

waveform data were inverted, geodetic measurements and surface rupture mapping revealed that the mainshock rupture propagated into multiple fault segments with different geometries. Figure 4.3 shows a source fault model for the foreshock and mainshock (Geospatial Information Authority of Japan, 2016). Three major fault segments are modeled to explain the surface deformation caused by the sequence. About 20-km-long Fault A1 released most of the elastic energy. This fault dips to the northwest at 60°, and the average slip was estimated at 4.1 m. There was a significant normal slip component in addition to the strike-slip component along this fault (Kato et al. 2016; Toda et al. 2016). Fault A2 is located within the caldera of Aso volcano, dipping southeast. The slip was predominantly right-lateral strike-slip.

The foreshock and mainshock caused severe ground motions in the epicentral area. Notably, severe seismic shaking of JMA seismic intensity scale 7 hit the Mashiki Town twice (Fig. 4.4). The peak acceleration recorded by surface instruments at Mashiki exceeded 15 m/s^2 (1500 gals). Horizontal component sensors recorded strong ground motions with dominant periods of 1–2 s, called "killer pulse." This pulse maximized the power for destroying wooden buildings due to resonance oscillations and was partly responsible for severe damage to many structures. Because the distribution of the destroyed houses in Mashiki by the mainshock are similar to the ones by the foreshock that was unrelated to the surface rupture, the majority of the structural damages were perhaps associated with the soil amplification due to near-surface ground condition (Fig. 4.5, Yamada 2017). However, some of the structural damages may be partly due to near-surface source properties (Naito et al. 2017; Irikura et al. 2020) and/or ground deformation, as several traces of surface ruptures appeared in the downtown Mashiki (Suzuki et al. 2018). Hisada et al. (2020) also pointed out structural damages near the surface ruptures along the entire fault zone.

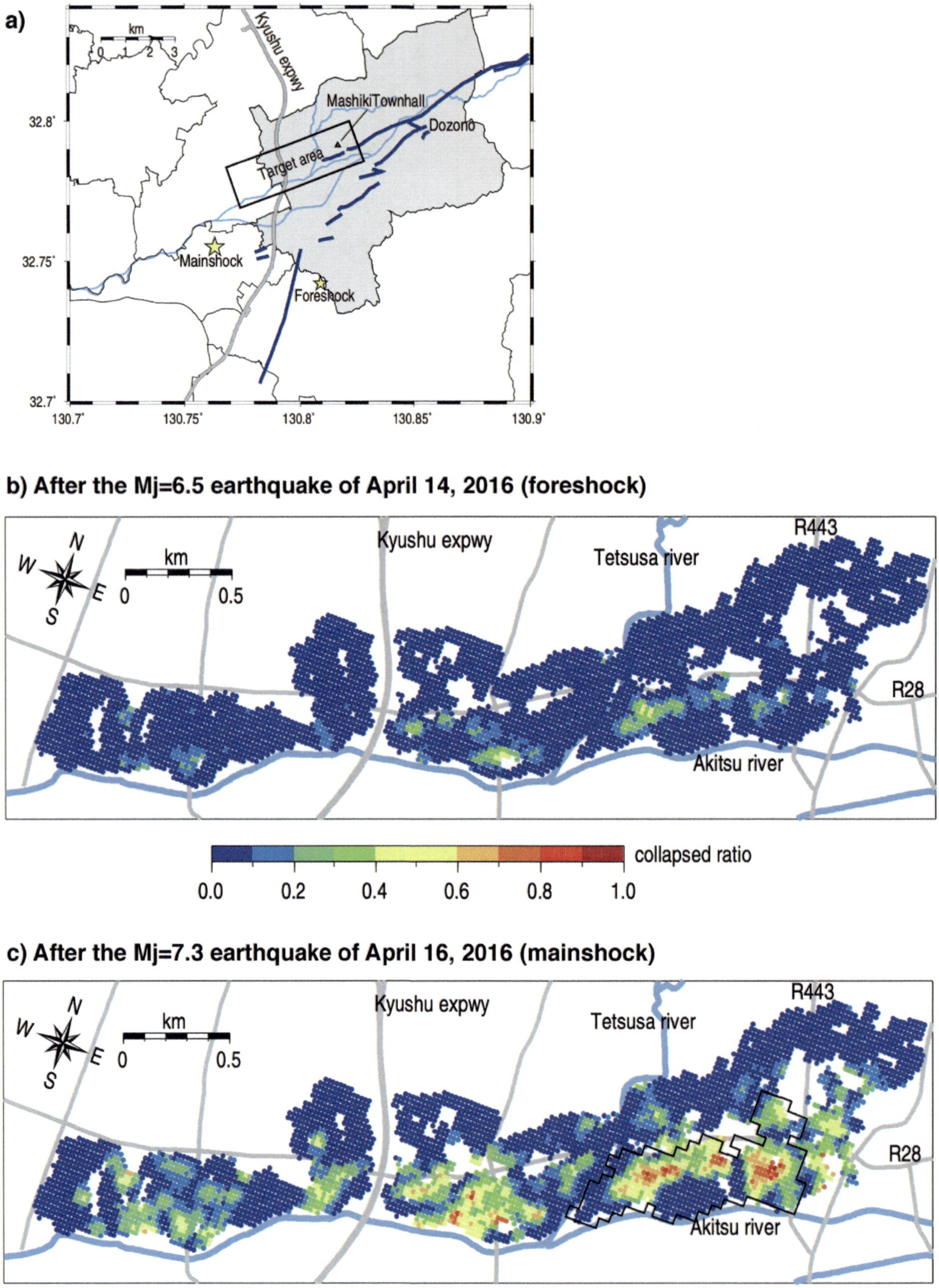

Fig. 4.5 Distribution of the collapsed buildings due to the 2016 Kumamoto earthquakes in the area along the Akitsu River, Mashiki Town (Yamada, 2017). Index map showing the spatial relationship with the surface rupture (**a**), collapsed rate associated with the foreshock (**b**), and collapsed rate identified after the mainshock (**c**)

References

Asano K, Iwata T (2016) Source rupture processes of the foreshock and mainshock in the 2016 Kumamoto earthquake sequence estimated from the kinematic waveform inversion of strong motion data. Earth Planets Space 68:147. https://doi.org/10.1186/s40623-016-0519-9

Geospatial Information Authority of Japan (2016) The 2016 Kumamoto Earthquake: Crustal deformation around the faults. http://www.gsi.go.jp/BOUSAI/H27-kumamoto-earthquake-index.html#3. Accessed 13 June 2021

Hisada Y, Tanaka S, Kaneda J, Teramoto A, Nakamura W, Murakami M, Masuzawa Y, Sakai S, Nakano K, Mori K, Kimoto K (2020) Investigation of building damage near surface fault rupture of the 2016 Kumamoto earthquake and countermeasures for active faults. J Jpn Assoc Earthq Eng 20:90–132

Irikura K, Kurahashi S, Matsumoto Y (2020) Extension of characterized source model for long-period ground motion in near-fault area. Pure Appl Geophys 177:2021–2047

Kato A, Nakamura K, Hiyama Y (2016) The 2016 Kumamoto earthquake sequence. Proc Jpn Acad Ser B Phys Biol Scis 92 (8):358–371

Kobayashi T (2017) Earthquake rupture properties of the 2016 Kumamoto Earthquake foreshocks (Mj6.5 and Mj6.4) revealed by conventional and multiple-aperture InSAR. Earth Planets Space 69:7. https://doi.org/10.1186/s40623-016-0594-y

Kobayashi T, Yarai H, Kawamoto S, Morishita Y, Fujiwara S, Hiyama Y (2018) Crustal deformation and fault models of the 2016 Kumamoto earthquake sequence: Foreshocks and mainshock. In: Freymueller J, Sánchez L (eds) International symposium on advancing geodesy in a changing world. International Association of Geodesy Symposia, vol 149. Springer, Heidelbeg, pp 193–200

Kubo H, Suzuki W, Aoi S, Sekiguchi H (2016) Source rupture processes of the 2016 Kumamoto, Japan, earthquakes estimated from strong-motion waveforms. Earth Planets Space 68:161. https://doi.org/10.1186/s40623-016-0536-8

Miyazawa M (2016) An investigation into the remote triggering of the Oita earthquake by the 2016 Mw 7.0 Kumamoto earthquake using full wavefield simulation. Earth Planets Space 68:205. https://doi.org/10.1186/s40623-016-0585-z

Naito S, Hao KX, Senna S, Saeki T, Nakamura H, Fujiwara H (2017) Investigation of damages in immediate vicinity of co-seismic faults during the 2016 Kumamoto earthquake. J Disaster Res 12:899–915

Nanjo KZ, Izutsu J, Orihara Y, Kamogawa M, Nagao T (2019) Changes in seismicity pattern due to the 2016 Kumamoto earthquakes identify a highly stressed area on the Hinagu fault zone. Geophys Res Lett. https://doi.org/10.1029/2019GL083463

Pollitz FF, Kobayashi T, Yarai H, Shibazaki B, Matsumoto T (2017) Viscoelastic lower crust and mantle relaxation following the 14–16 April 2016 Kumamoto, Japan, earthquake sequence. Geophys Res Lett 44:8795–8803. https://doi.org/10.1002/2017GL074783

Shimizu Y, Matsumoto S, Matsushima K, Aizawa H, Sido A, Yamashita H (2017) Seismic investigations to reveal three-dimensional structure of the fault zone, In: Ministry of Education, Culture, Sports, Science and Technology, and Kyushu University (eds) Report of the comprehensive survey of the 2016 Kumamoto earthquake. https://www.jishin.go.jp/main/chousakenkyuu/kumamoto_sogochousa/h28/h28kumamoto_sogochousa_3_2.pdf. Accessed 13 June 2021

Sugito N, Goto H, Kumahara Y, Tsutsumi H, Nakata T, Kagohara K, Matsuta N, Yoshida H (2016) Surface fault ruptures associated with the 14 April foreshock (Mj 6.5) of the 2016 Kumamoto earthquake sequence, southwest Japan. Earth Planets Space. https://doi.org/10.1186/s40623-016-0547-5

Suzuki Y, Watanabe M, Nakata T (2018) Surface earthquake faults in the urbanized area of Mashiki town associated with the 2016 Kumamoto Earthquake of Japan—Tectonic significance and impact on building damages. Active Fault Res 48:13–34

Toda S, Kaneda H, Okada S, Ishimura D, Mildon Z (2016) Slip-partitioned surface ruptures for the Mw 7.0 16 April 2016 Kumamoto, Japan, earthquake. Earth Planets Space. https://doi.org/10.1186/s40623-016-0560-8

Yamada M (2017) Damage islands in Mashiki town from the 2016 Kumamoto earthquakes. J JAEE (Jpn Assoc Earthquake Eng). https://doi.org/10.5610/jaee.17.5_38

Yoshida S (2016) Earthquakes in Oita triggered by the 2016 M7.3 Kumamoto earthquake. Earth Planets Space 68:176. https://doi.org/10.1186/s40623-016-0552-8

Part II

Detailed Description of the Surface Ruptures

Field Mapping Methods and Data Compilation Procedures of the Surface Ruptures

5

Yasuhiro Kumahara, Heitaro Kaneda, and Hiroyuki Tsutsumi

Abstract

In this chapter, we briefly described our field mapping methods of the 2016 surface ruptures, including the response of our group to the earthquake sequence, the slip measurement method, and the method of locating field observation points. We also provided the structure of this book. Detailed descriptions are given in the separate 12 chapters that correspond to different fault sections or areas in the 2016 surface ruptures.

Keywords

2016 Kumamoto earthquake • Surface rupture • Field mapping

Supplementary Information
The online version contains supplementary material available at (https://doi.org/10.1007/978-981-19-1150-7_5).
Note
The supplementary material of this book contains geographical information system (GIS) and spreadsheet files of the surface ruptures associated with the mainshock of the 2016 Kumamoto earthquake sequence

Y. Kumahara (✉)
Graduate School of Humanities and Social Sciences, Hiroshima University, 1-1-1, Kagamiyama, Higashi-Hiroshima, Hiroshima, 739-8524, Japan
e-mail: kumakuma@hiroshima-u.ac.jp

H. Kaneda
Department of Civil and Environmental Engineering, Chuo University, 1-13-27, Kasuga, Bunkyo-ku, Tokyo, 112-8551, Japan
e-mail: kaneda.00c@g.chuo-u.ac.jp

H. Tsutsumi
Department of Environmental Systems Science, Faculty of Science and Engineering, Doshisha University, 1-3, Tatara-Miyakodani, Kyotanabe, Kyoto, 610-0394, Japan
e-mail: htsutsum@mail.doshisha.ac.jp

Our primary field mapping of the surface ruptures started on April 15, 2016, after the 14 April Mj 6.5 (Mw 6.2) foreshock. Immediately after the 16 April Mj 7.3 (Mw 7.0) mainshock that appeared to accompany extensive surface faulting, we communicated with each other, set up a group mainly composed of researchers from Japanese universities, and began a systematic mapping campaign to cover the entire surface ruptures of the 2016 earthquake sequence. The concentrated field campaign continued to the end of May 2016, and after that each of our group members intermittently visited the affected area and conducted additional mapping until 2018.

In the field, we followed the surface ruptures and identified displaced and deformed artificial features that served as piercing points to accurately measure the lateral and vertical displacements. We used a variety of tools to measure the displacement, including measuring tapes, foldable rulers, leveling poles, hand levels, and laser distance meters. There were often few artificial features in the mountainous areas, and in that case, we could not measure the displacement amount accurately, especially that of the lateral slip. To record the geographical locations of the observation points, we principally used handheld GNSS devices. Since locations from the handy GNSS are associated with ~3–10 m horizontal errors depending on field and satellite conditions, we also mapped the precise locations of the surface ruptures by combining the GNSS readings and post-earthquake orthorectified aerial photographs provided by the Geospatial Information Authority of Japan. The aerial photos have a ~5-cm spatial resolution; therefore, we could readily identify and locate the surface ruptures on the images. We also used the 1–2-m-grid digital elevation models along the surface ruptures acquired by the Geospatial Information Authority of Japan to identify and measure coseismic and cumulative displacements across the faults. The data from those field mapping were gathered and compiled by Y. Kumahara.

Y. Kumahara et al. (eds.), *Surface Ruptures Associated with the 2016 Kumamoto Earthquake Sequence in Southwest Japan*, Advances in Geological Science, https://doi.org/10.1007/978-981-19-1150-7_5

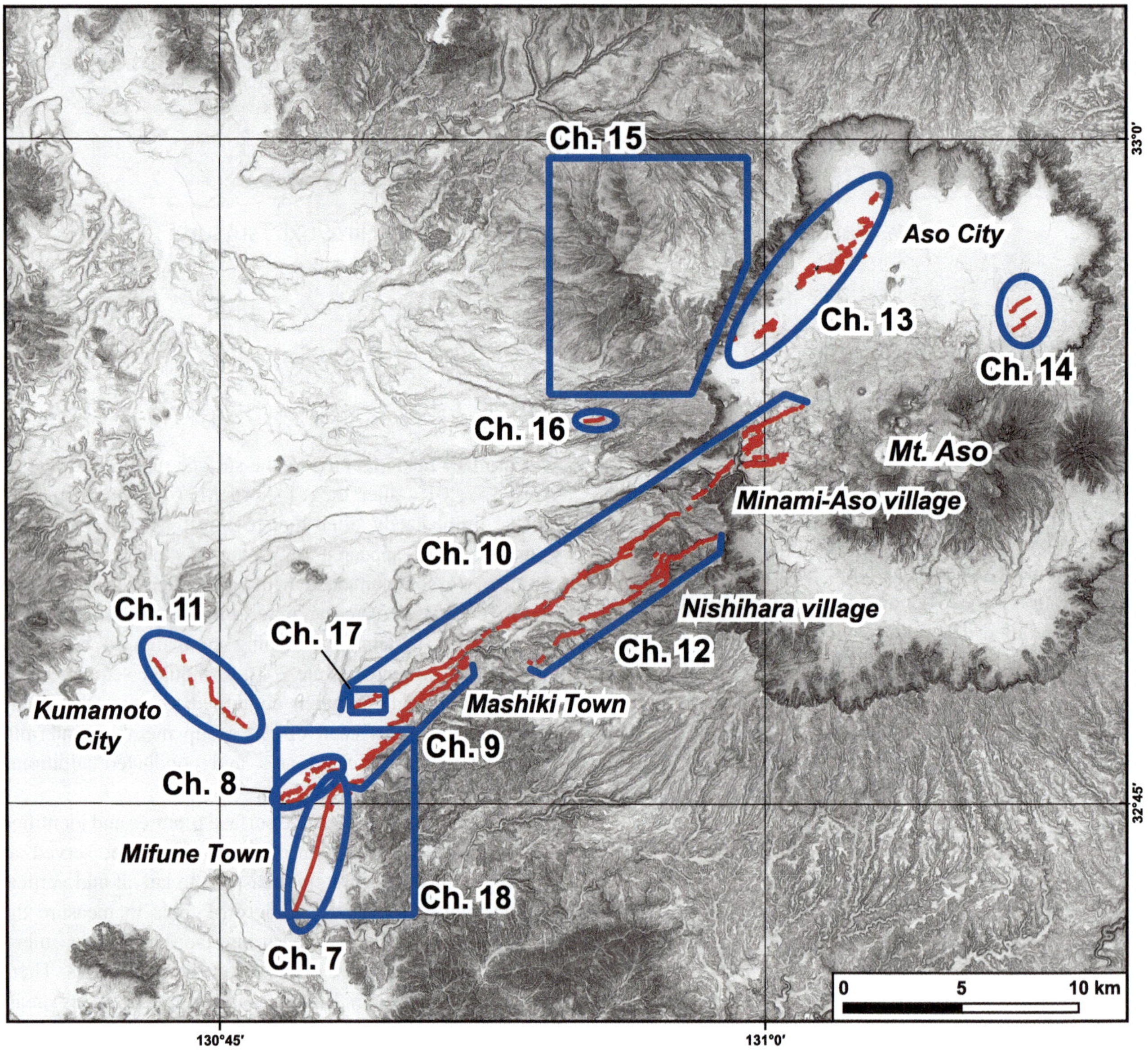

Fig. 5.1 Index map of the surface ruptures covered in each Chap. 7 through 18. Red lines indicate the surface rupture traces of the 2016 Kumamoto earthquake sequence as mapped by our group

In the following chapters, we first summarize general characteristics of the 2016 surface ruptures as mapped by our group in Chap. 6. We then describe the entire surface ruptures in detail by dividing them into 12 sections or areas chiefly by their geometric and geomorphic characteristics (Fig. 5.1):

Chapter 7: Surface Ruptures of the Shirahata–Oike Section
Chapter 8: Surface Ruptures along the Kita-Amagi Fault Zone
Chapter 9: Surface Ruptures along the Southern Part of the Futagawa Fault
Chapter 10: Surface Ruptures along the Central-Northern Part of the Futagawa Fault
Chapter 11: Surface Ruptures in the Downtown of Kumamoto City
Chapter 12: Surface Ruptures and Tectonic Geomorphology along and around the Idenokuchi Fault
Chapter 13: Surface Ruptures in the Northwestern Part of the Inner Aso Caldera
Chapter 14: Surface Ruptures in the Northeastern Part of the Inner Aso Caldera
Chapter 15: Surface Ruptures in the Northwest of the Outer Aso Caldera
Chapter 16: Surface Ruptures along the Western Part of the Bungo-Kaido Road
Chapter 17: Surface Ruptures in Mashiki Town: Tectonic Significance and Building Damage

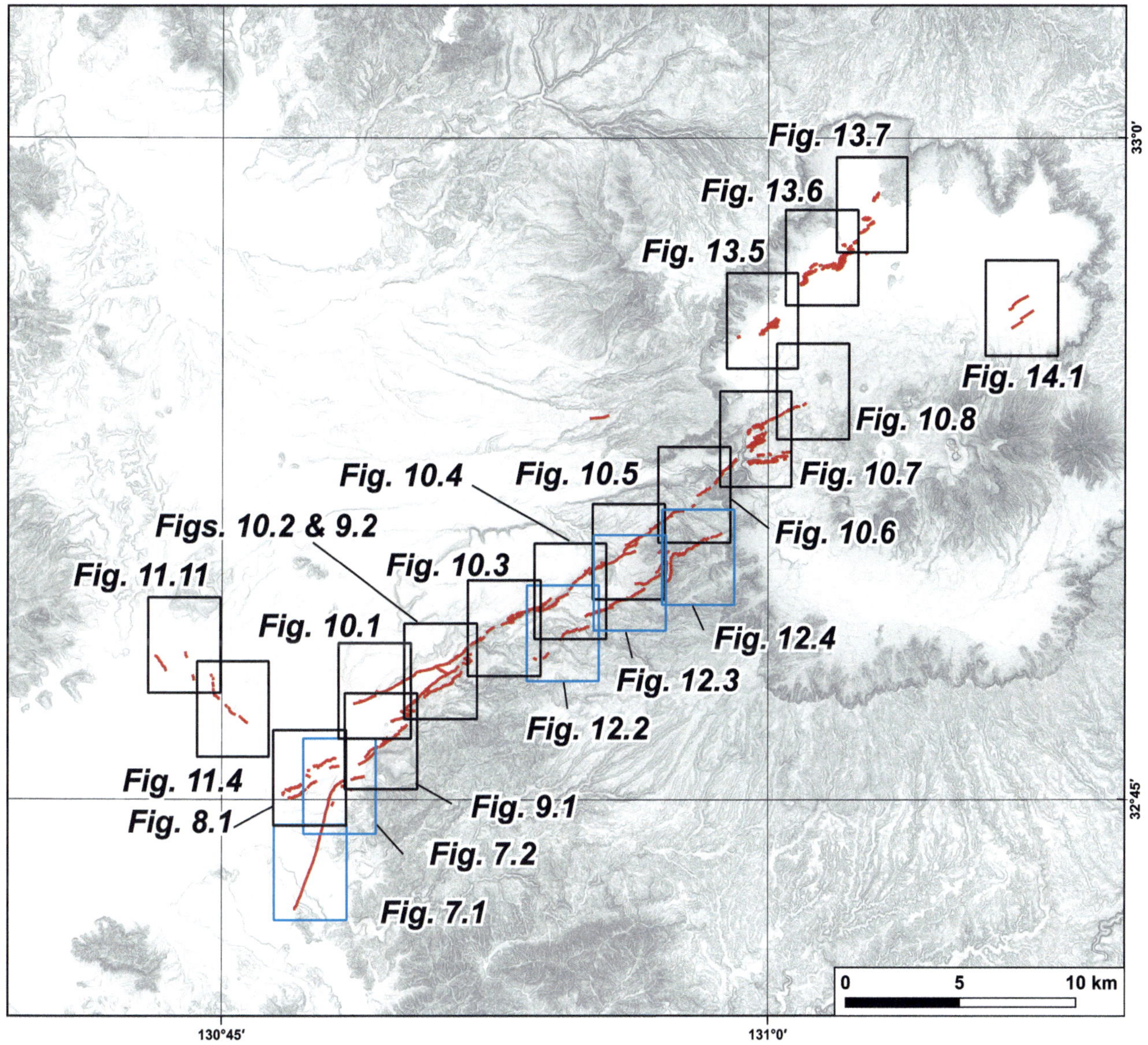

Fig. 5.2 Index map of the large-scale surface rupture maps shown in the following chapters. Red lines indicate the surface rupture traces of the 2016 Kumamoto earthquake sequence as mapped by our group

Chapter 18: Surface Ruptures Accompanied with the Largest Foreshock.

In each chapter from Chap. 7 through 14, large-scale surface rupture maps are given with a common format, in which observation and slip measurement points are shown with the sense and amount of fault slips. The index map for those large-scale maps is shown in Fig. 5.2.

For readers' convenience, we also provide digital data of the surface ruptures as the electronic supplementary material of this book. The data include line shape files of the surface rupture traces, point shape files of the slip measurement points, and the Excel spreadsheet file of the slip measurement points.

General Characteristics of the Surface Ruptures of the 2016 Kumamoto Earthquake Sequence

6

Yasuhiro Kumahara, Heitaro Kaneda, and Hiroyuki Tsutsumi

Abstract

In this chapter, we compiled and summarized the information of the surface ruptures accompanied with the 2016 Kumamoto earthquake sequence to provide an overview of the surface ruptures and introduction to the following chapters. The primary surface ruptures appeared principally along the previously mapped active faults, such as the Futagawa, Hinagu, and Idenokuchi faults, but many scattered and isolated surface breaks were also found around and away from the main surface ruptures. The end-to-end distance of the main surface ruptures excluding the isolated surface breaks was ~30 km, but the total length of the mapped surface ruptures reached as long as ~87 km. The maximum right-lateral offset was 225 cm on the Futagawa fault, whereas the maximum vertical displacement of 190 cm down to the west was found on the unmapped section of the Idenokuchi normal fault.

Supplementary Information
The online version contains supplementary material available at (https://doi.org/10.1007/978-981-19-1150-7_6).
Note
The supplementary material of this book contains geographical information system (GIS) and spreadsheet files of the surface ruptures associated with the mainshock of the 2016 Kumamoto earthquake sequence

Y. Kumahara (✉)
Graduate School of Humanities and Social Sciences, Hiroshima University, 1-1-1, Kagamiyama, Higashi-Hiroshima, Hiroshima, 739-8524, Japan
e-mail: kumakuma@hiroshima-u.ac.jp

H. Kaneda
Department of Civil and Environmental Engineering, Chuo University, 1-13-27, Kasuga, Bunkyo-ku, Tokyo, 112-8551, Japan
e-mail: kaneda.00c@g.chuo-u.ac.jp

H. Tsutsumi
Department of Environmental Systems Science, Faculty of Science and Engineering, Doshisha University, 1-3, Tatara-Miyakodani, Kyotanabe, Kyoto, 610-0394, Japan
e-mail: htsutsum@mail.doshisha.ac.jp

Keywords

2016 Kumamoto earthquake • Surface rupture • Futagawa fault

After the 2016 Kumamoto earthquake sequence, in particular the Mj 7.3 (Mw 7.0) mainshock on 16 April, many researchers rushed to the epicentral area and confirmed at many locations that the causative fault ruptured the surface during the event. Several research groups attempted systematic mapping of the surface ruptures, including our group, and some of their results have already been published in research journals and maps. A team from the Geological Survey of Japan, AIST, promptly presented the traces and slip distribution of the main surface ruptures (Shirahama et al. 2016). Toda et al. (2016) focused on the northern part of the Futagawa fault and the Idenokuchi fault and described in detail the surface ruptures along those fault sections. Kumahara et al. (2017) and Suzuki et al. (2017) showed the 2016 surface rupture traces on 1:25,000-scale active fault maps, which are principally based on an earlier version of the surface rupture data described in this book. Incorporating new information and discoveries since then, the following chapters present more detailed maps of all the surface ruptures, many on-site photographs taken immediately after the earthquake, and detailed explanations of the features and displacements of the surface ruptures.

The primary surface ruptures of the 2016 Kumamoto earthquake sequence appeared principally along the previously mapped active faults, such as the Futagawa, Hinagu, and Idenokuchi faults, but many scattered and isolated surface breaks were also found around and away from the main surface ruptures (Fig. 6.1). The end-to-end distance of the surface ruptures excluding the isolated surface breaks was ~30 km, but the total length of the mapped surface ruptures reached as long as ~87 km. In Fig. 6.2, we show detailed

Y. Kumahara et al. (eds.), *Surface Ruptures Associated with the 2016 Kumamoto Earthquake Sequence in Southwest Japan*, Advances in Geological Science, https://doi.org/10.1007/978-981-19-1150-7_6

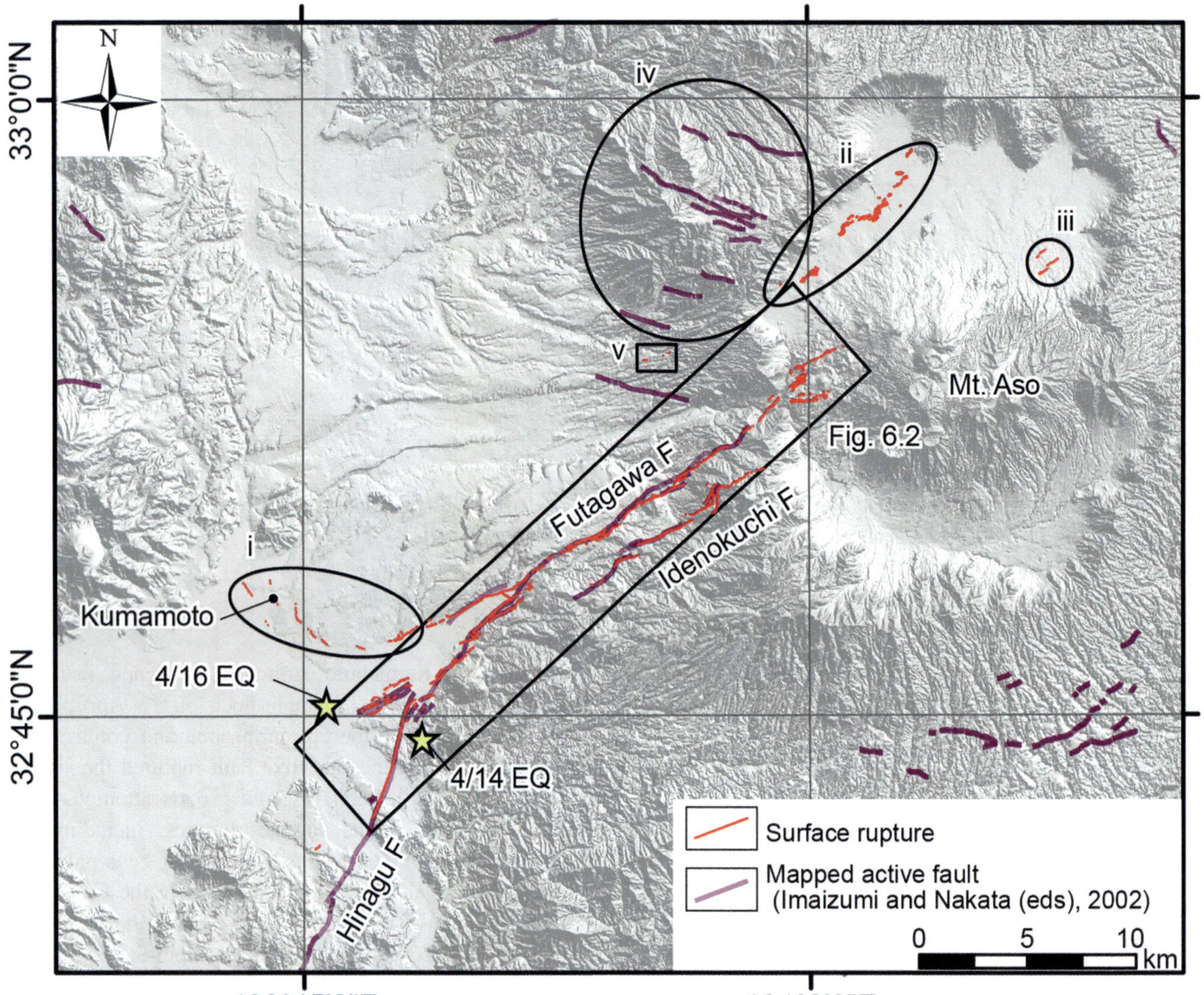

Fig. 6.1 Surface ruptures of the 2016 Kumamoto earthquake sequence (red lines) and the mapped active faults in and around the epicentral area from Imaizumi and Nakata (2002) (purple lines)

slip distribution along individual sections of the main surface ruptures. The maximum right-lateral offset was 225 cm on the Futagawa fault, whereas the maximum vertical displacement of 190 cm down to the west was found on the unmapped section of the Idenokuchi normal fault.

The ~7-km-long southernmost segment of the ruptures from Shirahata to Oike (segment A in Fig. 6.2; Chap. 7) is quite simple and straight. This segment strikes N20°E, rotated clockwise from the general strike of the entire surface ruptures (N50–60°E), and ruptured a part of the NNE-trending Hinagu fault. The slip is predominantly right-lateral strike-slip with 50–65 cm displacement between 2.5 and 6 km from the southern end and tapers to both ends.

North of segment A, the surface ruptures consist of short rupture segments less than 2 km long (segments B, C, D, and E; Chap. 8). Segments B and C are subparallel and strike ~N60°E, and are reactivation of the previously mapped Kita-Amagi fault zone (Imaizumi and Nakata 2002). The vertical displacement on segment B is less than 30 cm up on the southeast, while the right-lateral offset is less than 20 cm. We (our group) observed less than 10 cm vertical displacement up on the northwest with no discernible strike-slip along segment C. Segments B and C bound the northern and southern margins of the preexisting elongated depression, consistent with the coseismic vertical displacement on the two ruptures. Segments D and E also strike ~N60°E. The right-lateral offset on segment D is less than 20 cm with a vertical displacement no more than 10 cm up on the southeast. We measured a right-lateral offset of ~20 cm on segment E.

Surface ruptures of segments F, G, H, Ia, Ib, J, and K appeared along the northwestern foot of the mountains to the

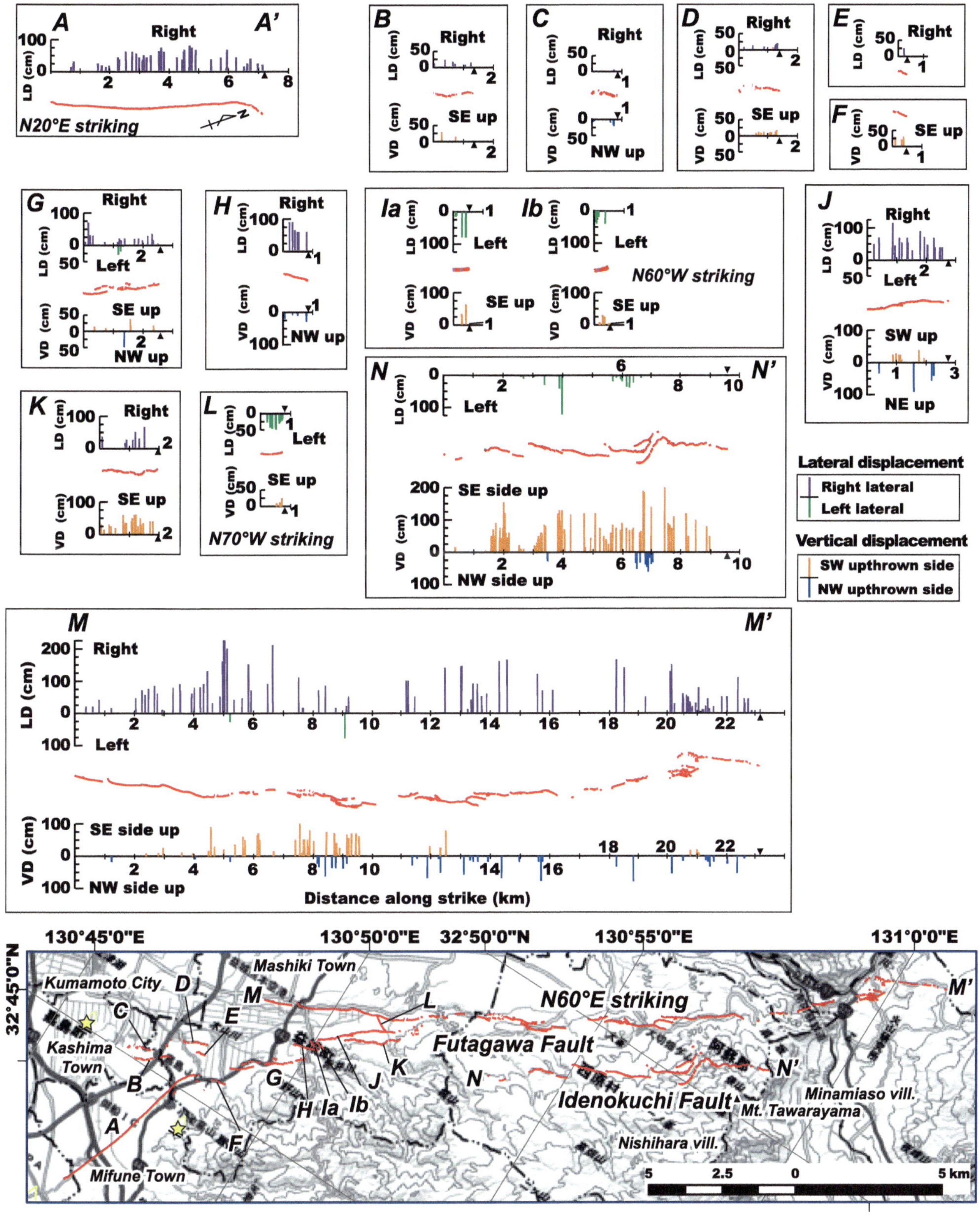

Fig. 6.2 Distribution of the lateral and vertical displacements on each segment of the surface ruptures of the 2016 Kumamoto earthquake sequence

south and extended for a length of 7.5 km (Chap. 9). These segments strike ~50°E, except for ~N80°E trending right-lateral segment H and the conjugate left-lateral segments Ia and Ib (~N60°W). Segments F, G, H, J, and K overlap with a left-stepping *en echelon* pattern. We measured less than 30 cm vertical displacement up on the southeast with no lateral offset on segment F. The vertical displacement along segment G reached 70 cm, while the right-lateral offset was less than 30 cm. The right-lateral offset was 60–90 cm with less than 30 cm up-to-the-northwest vertical displacement along segment H. Segments Ia and Ib are subparallel and ~500 m long, and conjugate to the right-lateral faults. The maximum left-lateral offset was 80 cm for segment Ia and 40 cm for segment Ib, while the vertical displacement was up on the southwest for both segments. Subparallel segments J and K are 2.5 km long and 2 km long, respectively. The right-lateral offset on segment J was 90–115 cm in the middle, decreasing to 30–40 cm at both ends. The direction of vertical displacement on segment J frequently changes along the strike, indicating that the vertical displacement was secondary to the predominant strike-slip. The vertical displacement along segment K was predominantly up on the southeast and reached ~60 cm in the middle part. The measured right-lateral offsets along segment K were 10–65 cm. It seems that slip partitioning occurred between segments J and K based on the slip pattern and distribution. Near the northeastern end of segment J, segment L obliquely cut across an alluvial plain north of segment J and extended for 800 m with a strike of ~70°W, forming a conjugate surface rupture. The maximum left-lateral offset was ~40 cm along segment L.

Segment M is a ~23-km-long main rupture segment of the 2016 earthquake sequence that coincides with the main part of the Futagawa fault (Chap. 10). It is largely continuous with occasional left steps, except for the locations where we could not trace the ruptures due to thick vegetation in the mountainous areas. In the northeastern part, segment M cut across the caldera wall of Aso Volcano and extended into the caldera. However, previous maps did not show any active faults inside the caldera. The slip is predominantly right-lateral with two peaks of the lateral displacement (Fig. 6.2). At the southern peak, the right-lateral offset reached 225 cm at 4.8 km from the south end, the largest measurement along the entire surface ruptures of the Kumamoto earthquake sequence. At the northern peak, the amount of right-lateral slip was 100–140 cm between 12 and 18.5 km from the south end, and it gradually tapered to the north end. The direction of vertical displacement along segment M changed systematically. Up to ~8 km from the south end, the up-thrown side was southeast with the maximum displacement of 80 cm, while it was predominantly northwest from a distance of ~12 km up to the northern end with the similar maximum displacement of 80 cm. This is consistent with the sense of minor vertical slip expected along a pure right-lateral fault. Distributed and scattered surface ruptures were also found around the western termination of segment M in the urbanized area of Mashiki Town (not shown in Fig. 6.2; Suzuki et al. 2018; Chap. 18).

Left-stepping surface ruptures of segment N appeared along the Idenokuchi fault for ~10 km (Chap. 12). The general strike of the segment was ~N60°E, but two subparallel traces at ~7 km from the southern end of the segment trended almost north–south. Up-to-the-southeast normal faulting was dominant, but there were several traces of up-to-the-northwest antithetic surface faulting. The amount of vertical displacement is generally 80–100 cm, but it reaches 190 cm in the northern half of the segment. Interestingly, left-lateral offsets were measured along the segment N in many places. Normal faulting on segment N was interpreted as the result of slip partitioning between the Futagawa and Idenokuchi faults (Toda et al. 2016).

Figure 6.3 showed the distribution of summed right-lateral slip along the general strike of the Futagawa fault (N60°E). The displacements on the subparallel faults were added to show the overall right-lateral displacement across the fault zone. The figure reveals that the overall slip increased drastically from ~0.5 m to > 2 m at ~ 8 km from the mainshock epicenter. The offset amount exceeded 2 m in the section between ~8 and ~14 km, reaching the maximum value of ~2.5 m at ~11 km. The slip then decreased gradually down to ~1 m at ~18 km, but again exceeded 1.5 m from ~18 to ~27 km.

Other surface ruptures accompanied with the 2016 Kumamoto earthquake sequence away from the main ruptures were observed in areas i to v (Fig. 6.1). In area i, open cracks appeared intermittently along the newly identified, NW-trending flexure scarps down to the southwest on the ~90 ka Aso-4 pyroclastic flow surface (Goto et al., 2017; Chap. 11). The tilting of artificial features to the west on the order of several centimeters indicated the growth of the tectonic scarp.

In area ii, numerous short open cracks appeared for ~8 km inside the northwest of the Aso caldera floor (Chap. 13). The ruptures consist of several parallel and subparallel sharp scarps by normal faulting that commonly form elongated fault depressions and horsts between those scarps. Along the surface traces, paddy fields were displaced vertically up to 2 m, and some were associated with ~30 cm right-lateral horizontal displacement.

Two short surface ruptures were found in area iii in the northeastern part of the caldera floor (Ishimura et al., 2021; Chap. 14). The ~1-km-long north strand showed a slightly sinuous trace with a small amount of right-lateral offset (<5 cm) and vertical displacement (<10 cm). We identified broad extensional deformation on the ~2-km-long south strand, such as open cracks on paved roads.

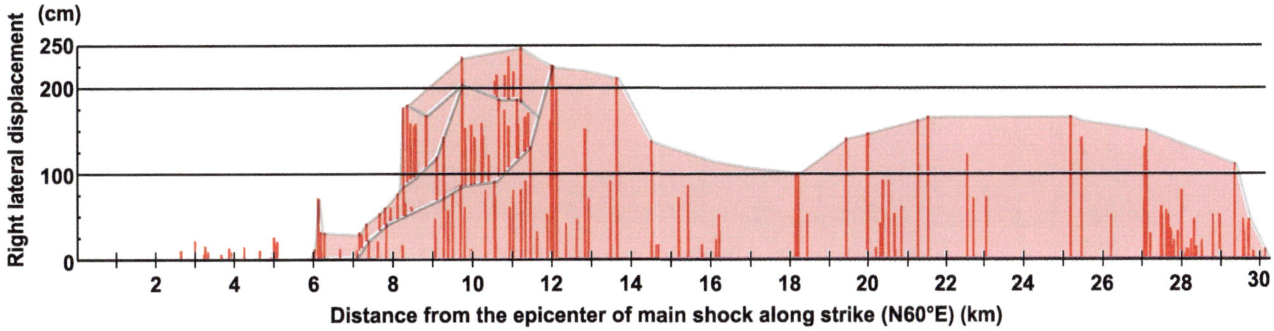

Fig. 6.3 Distribution of summed right-lateral displacement projected to the general strike of the Futagawa fault (N60°E), where the displacements on the subparallel faults were added to show the overall displacement across the fault zone

In the northwestern outer rim of Aso caldera (area iv), Fujiwara et al. (2016) mapped the "linear surface ruptures (LSR)" based on the SAR interferometric analysis, 5–20 km away from the Futagawa fault (Chap. 15). We mapped a lot of scattered small surface ruptures along the LSR in the field. These surface ruptures showed normal faulting with a small slip less than a few tens of centimeters.

In area v, surface ruptures were found at six localities on the known active faults along the Bungo-kaido road (Chap. 16). The offset was right-lateral strike-slip on the order of several centimeters.

In addition to the extensive surface ruptures associated with the 16 April Mj 7.3 mainshock, the 14 April Mj 6.5 foreshock accompanied minor surface ruptures along parts of segments C, D, and G (Sugito et al. 2016; Chap. 18).

References

Fujiwawa S, Yarai H, Kobayashi T, Morishita Y, Nakano T, Miyahara B, Nakai H, Miura Y, Ueshiba H, Kakiage Y, Une H (2016) Small-displacement linear surface ruptures of 2016 Kumamoto earthquake sequence detected by ALOS-2 SAR interferometry. Earth Planets Space 68:160. https://doi.org/10.1186/s40623-016-0534-x

Goto H, Tsutsumi H, Toda S, Kumahara Y (2017) Geomorphic features of surface ruptures associated with the 2016 Kumamoto earthquake in and around the downtown of Kumamoto City, and implications on triggered slip along active faults. Earth Planets Space 69:26. https://doi.org/10.1186/s40623-017-0603-9

Imaizumi T, Nakata T (eds) (2002) Digital active fault map of Japan. University of Tokyo Press, Tokyo

Ishimura D, Tsutsumi H, Toda S, Fukushima Y, Kumahara Y, Takahashi N, Ichihara T, Takada K (2021), Repeated triggered ruptures on a distributed secondary fault system: an example from the 2016 Kumamoto earthquake, southwest Japan. Earth, Planets Space 73:39. https://doi.org/10.1186/s40623-021-01371-x

Kumahara Y, Okada S, Kagohara K, Kaneda H, Goto H, Tsutsumi H (2017) Active fault map 'Kumamoto" [revised edition]. Geospatial Information Authority of Japan D1-No.868, Geospatial Information Authority of Japan, Tokyo

Shirahama Y, Yoshimi M, Awata Y, Maruyama T, Azuma T, Miyashita Y, Miyakawa A (2016) Characteristics of the surface ruptures associated with the 2016 Kumamoto earthquake sequence, central Kyushu, Japan. Earth Planets Space 68:191. https://doi.org/10.1186/s40623-016-0559-1

Sugito N, Goto H, Kumahara Y, Tsutsumi H, Nakata T, Kagohara, K, Matsuta N, Yoshida H (2016) Surface fault ruptures associated with the 14 April foreshock (Mj 6.5) of the 2016 Kumamoto earthquake sequence, southwest Japan. Earth Planets Space 68:170. https://doi.org/10.1186/s40623-016-0547-5

Suzuki Y, Ishimura D, Kumaki Y, Kumahara Y, Chida N, Nakata T, Nakano T (2017) Active fault map 'Aso". Geospatial Information Authority of Japan D1-No.868, Geospatial Information Authority of Japan, Tokyo

Suzuki Y, Watanabe M, Nakata T (2018) Surface earthquake faults in the urbanized area of Mashiki town associated with the 2016 Kumamoto Earthquake of Japan—tectonic significance and impact on building damages. Active Fault Res 48:13–34

Toda S, Kaneda H, Okada S, Ishimura D, Mildon Z (2016) Slip-partitioned surface ruptures for the Mw7.0 16 April 2016 Kumamoto, Japan, earthquake. Earth Planets Space 69:188. https://doi.org/10.1186/s40623-016-0560-8

Surface Ruptures of the Shirahata–Oike Section

7

Yasuhiro Kumahara, Tatsuya Ishiyama, Nobuhisa Matta, Kyoko Kagohara, Daisuke Hirouchi, and Satoshi Ishiguro

Abstract

The ~7-km-long N20°E-trending southernmost section of the 2016 surface rupture extended from Shirahata to Oike and is characterized by a relatively simple and straight trace. This section cuts across alluvial plains and fluvial terrace surfaces and closely coincided with the previously mapped active fault trace. The section seemed to rupture the northernmost part of the NNE-trending Hinagu fault, or a transition zone between the Hinagu fault and the NE-trending Futagawa fault. Right-lateral offset was commonly observed along this section of the surface rupture, with offsets ranging from a few to 65 cm. In contrast, vertical displacement was largely negligible.

Keywords

Surface rupture • Hinagu fault • 2016 Kumamoto earthquake

Supplementary Information
The online version contains supplementary material available at (https://doi.org/10.1007/978-981-19-1150-7_7).
Note
The supplementary material of this book contains geographical information system (GIS) and spreadsheet files of the surface ruptures associated with the mainshock of the 2016 Kumamoto earthquake sequence

Y. Kumahara (✉)
Graduate School of Humanities and Social Sciences, Hiroshima University, 1-1-1, Kagamiyama, Higashi-Hiroshima, Hiroshima, 739-8524, Japan
e-mail: kumakuma@hiroshima-u.ac.jp

T. Ishiyama
Earthquake Research Institute, The University of Tokyo, 1-1-1, Yayoi, Bunkyo-ku, Tokyo, 113-0032, Japan
e-mail: ishiyama@eri.u-tokyo.ac.jp

N. Matta
Graduate School of Education, Okayama University, 3-1-1, Tsushima-naka, Kita-ku, Okayama, Okayama, 700-8530, Japan
e-mail: matta@okayama-u.ac.jp

K. Kagohara
Faculty of Education, Yamaguchi University, 1677-1, Yoshida, Yamaguchi, Yamaguchi, 753-8513, Japan
e-mail: k-kago@yamaguchi-u.ac.jp

D. Hirouchi
Department of Geography, Faculty of Education, Shinshu University, 6-ro, Nishi-Nagano, Nagano, Nagano, 380-8544, Japan
e-mail: hirokuma@shinshu-u.ac.jp

S. Ishiguro
Department of Humanities and Social Sciences, Faculty of Law and Letters, Ehime University, 3, Bunkyo-cho, Matsuyama, Ehime, 790-8577, Japan
e-mail: ishiguro.satoshi.uv@ehime-u.ac.jp

The ~7-km-long southernmost section of the 2016 surface rupture from Shirahata to Oike (Shirahata–Oike section) was characterized by a relatively simple and straight trace (Figs. 7.1 and 7.2). This section extended across alluvial plains and fluvial terrace surfaces and closely coincided with the previously mapped active fault trace (Ikeda et al. 2001). The strike of this section (N20°E) was rotated clockwise from the general trend of the entire surface rupture (N50–60° E), suggesting that this section ruptured a part of the NNE-trending Hinagu fault rather than the NE-trending Futagawa fault, or a transition zone between the two major faults. Right-lateral offset was commonly observed along this section of the surface rupture, with offsets ranging from a few to 65 cm. In contrast, vertical displacement was negligible except for Locs. 005 and 007 in the south.

We were able to trace the surface breaks southward to Loc. 001 at Shirahata, where the wall of an irrigation channel was right-laterally offset 2 cm. After the earthquake, Shirahama et al. (2020) conducted a trench survey at Loc. 001 and found several down-to-the-west near-vertical faults that clearly displace and deform Holocene back marsh, channel and fan deposits. Interestingly, the vertical separation associated with the penultimate faulting event was over

Y. Kumahara et al. (eds.), *Surface Ruptures Associated with the 2016 Kumamoto Earthquake Sequence in Southwest Japan*, Advances in Geological Science, https://doi.org/10.1007/978-981-19-1150-7_7

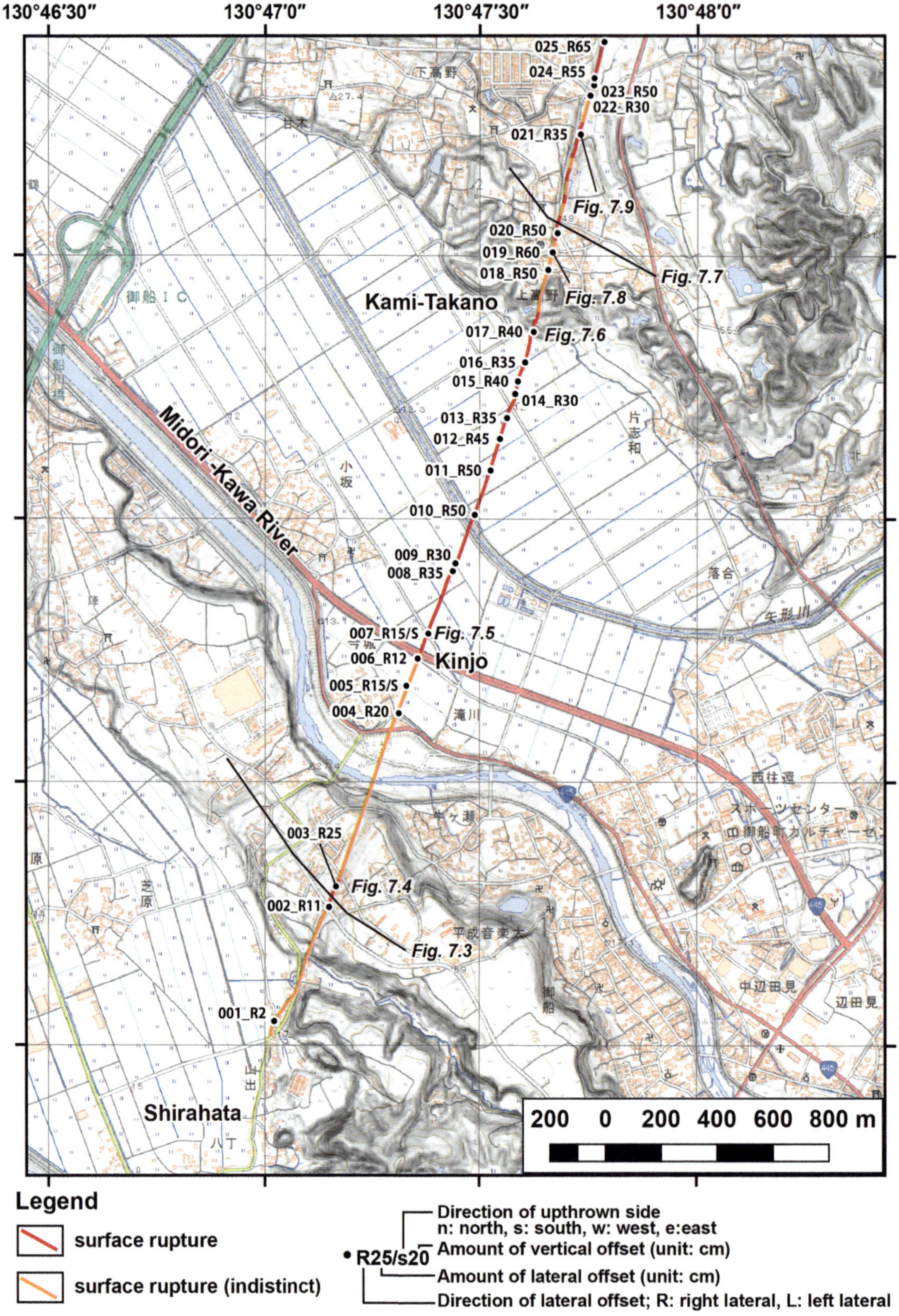

Fig. 7.1 Detailed surface rupture map of the southern part of the Shirahata–Oike section. Strike-slip and vertical displacements at each locality are shown. The background is a half-transparent topographic map of Geospatial Information Authority of Japan (GSI) underlain by a slope map produced from GSI 5-m grid digital elevation models (DEMs)

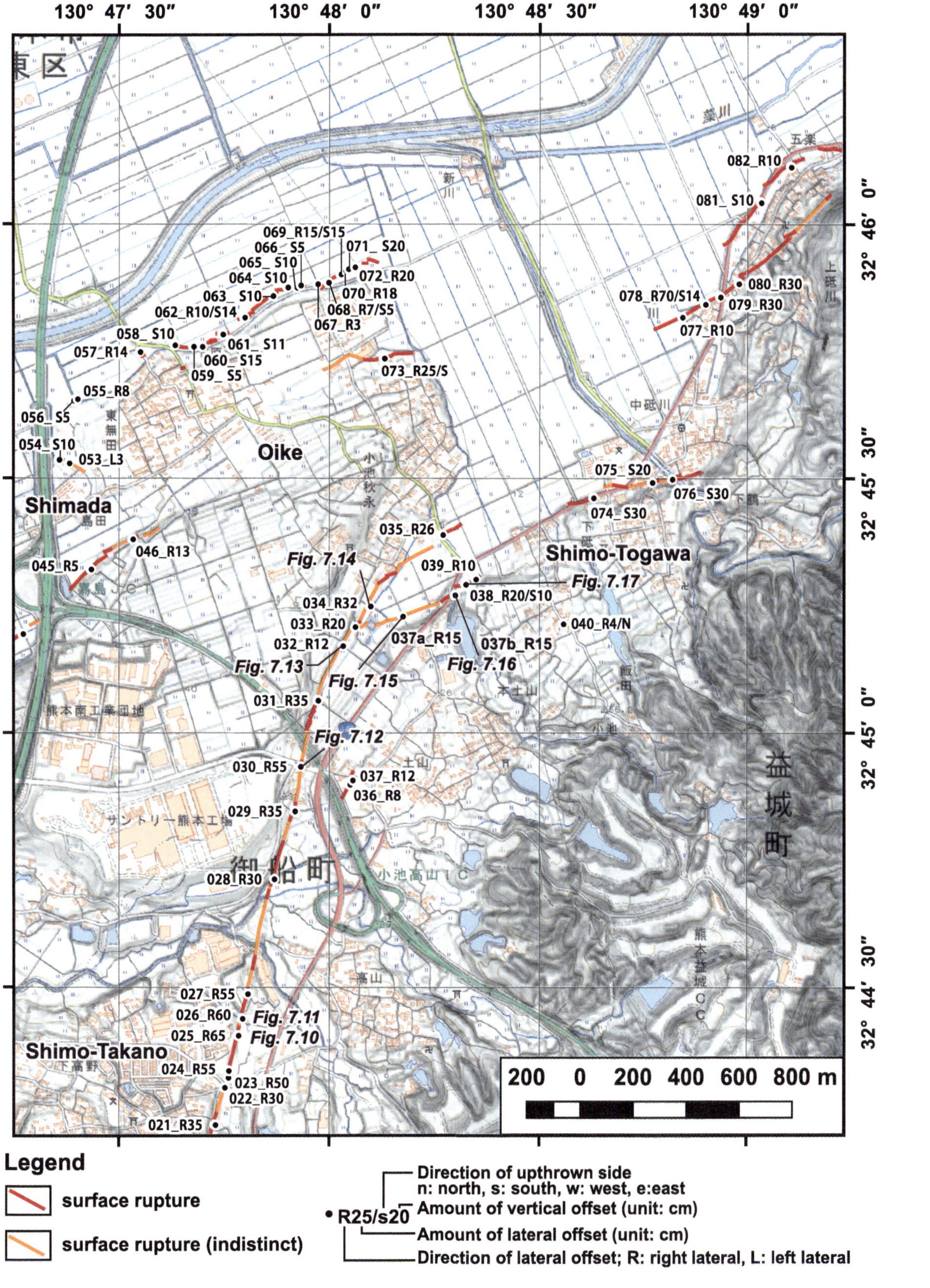

Fig. 7.2 Surface ruptures in the northern part of the Shirahata–Oike section. Strike-slip and vertical displacements at each locality are shown. The base map is the same as in Fig. 7.1

20 cm, in contrast to no visible vertical displacement during the 2016 earthquake.

We identified the surface ruptures along the base of the ~7-m-high west-facing fault scarp on a fluvial terrace between Locs. 002 and 003 (Fig. 7.3). Vegetable-field furrows were right-laterally offset ~25 cm with no visible vertical displacement at Loc. 003 (Fig. 7.4). Since the 2016 surface rupture did not appear to contribute to the growth of the fault scarp, the earlier surface faulting events would have been associated with a substantial amount of vertical displacement. From Locs. 003 to 006, the surface rupture was not continuous and was only identified at Locs. 004 and 005.

On the alluvial plain on the right (northern) bank of the Midori-kawa River, the surface rupture ran across rice paddy fields from Loc. 006 at Kinjo to Loc. 017 at Kami-Takano. Based on the offset measurements of straight footpaths and roads in the rice paddy field, the amount of right-lateral

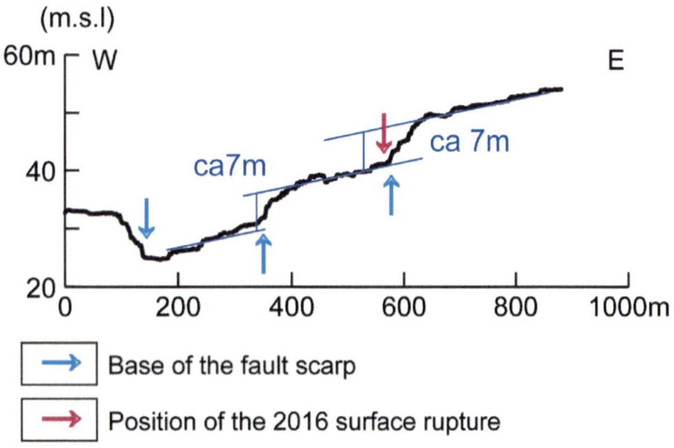

Fig. 7.3 Topographic profile across the fault scarps on a fluvial terrace surface between Locs. 002 and 003 based on 10-m grid DEMs of GSI. Blue arrows indicate the base of the fault scarps, and red arrow indicates the position of the 2016 surface rupture. See Fig. 7.1 for the location and extent of the profile

Fig. 7.4 Right-lateral offset of vegetable-field furrows at Loc. 003. Looking to the east. White arrows indicate a small ridge between the furrows as a reference of offset. Red arrows indicate the 2016 surface rupture. The amount of right-lateral offset was ~25 cm

Fig. 7.5 Right-lateral offset (~15 cm) with up-to-the-north vertical displacement of the centerline on the paved road at Loc. 007. White arrows indicate a straight orange line on the paved road as a reference of offset. Red lines show the 2016 surface rupture

offset increased northward from 12–15 cm at Locs. 006 and 007 (Fig. 7.5) to 35–50 cm at Locs. 010–017 (Fig. 7.6).

On a terrace surface to the north, there was a ~6-m-high uphill-facing (east-facing) fault scarp (Fig. 7.7), suggesting that the direction of cumulative vertical slip changed somewhere on the alluvial plain to the south. The 2016 surface rupture appeared along a preexisting linear depression at the scarp base from Locs. 018 to 020 for ~300 m. The right-lateral offset ranged from 50 to 60 cm (Fig. 7.8) with a left-stepping en-echelon structure (Fig. 7.9) with no visible vertical offset. At Loc. 023, Lin et al. (2017) conducted a trenching survey across the surface rupture after the earthquake and exposed several near-vertical faults cutting alluvial sediments. They identified three events,

Fig. 7.6 Right-lateral offset (~40 cm) of the edge of the paved road at Loc. 017. Red arrows indicate the fault trace

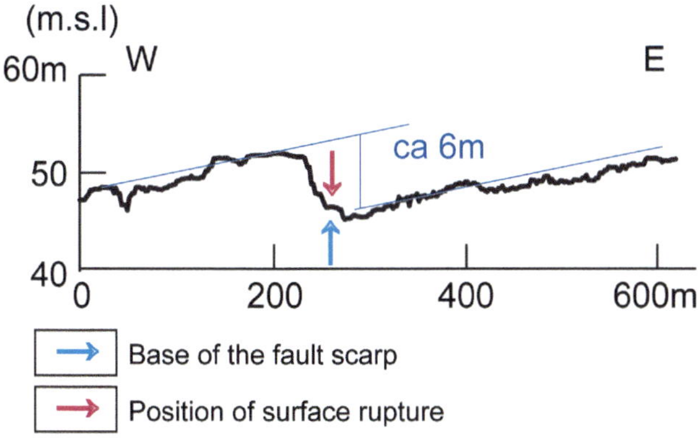

Fig. 7.7 Topographic profile across the fault scarp on a fluvial terrace surface close to Loc. 020 based on 10-m grid DEMs of GSI. Blue and red arrows indicate the base of the fault scarp and the position of the 2016 surface rupture, respectively. See Fig. 7.1 for the location and extent of the profile

Fig. 7.8 Right-lateral offset (~60 cm) of the white line on the paved road at Loc. 019. Red arrows indicate the 2016 surface rupture

including the 2016 earthquake, for the last 2000 years. At Loc. 025, we observed 65-cm right-lateral offset, the maximum slip on this section (Fig. 7.10). A stream dissecting the terrace surface flowed westward and crossed the fault trace at Loc. 026. As the sense of cumulative vertical slip was up-on-the-west, there was a depression on the eastern (upstream) side of the fault trace with a narrow outlet gouge cutting through the western (downstream) side. The 2016 event, however, did not contribute to the growth of the SE-facing fault scarp as a pure right-lateral strike-slip with

Fig. 7.9 Surface rupture with a left-stepping *en-echelon* structure at Loc. 021

Fig. 7.10 Right-lateral offset of footpath in the paddy field at Loc. 025. The amount of offset was 65 cm, the maximum slip in the Shirahata–Oike section. Red arrows indicate the 2016 surface rupture

Fig. 7.11 Right-lateral offset (~60 cm) of the edge of a paddy field and tractor ruts at Loc. 026. Shimokawa and Kinugasa (1999) conducted a trench survey here. Red arrows indicate the 2016 surface rupture

Fig. 7.12 Right-lateral offset of the white line on the paved road at Loc. 030 The amount of the offset was 55 cm. Red arrows indicate the 2016 surface rupture

Fig. 7.13 Surface rupture at the base of the east-facing fault scarp at Loc. 032

Fig. 7.14 Right-lateral offset of a drainage ditch of the paved road at Loc. 034. The amount of the offset was 32 cm. Red arrows indicate the 2016 surface rupture. The scale is 50 cm long

no vertical offset was recognized in 2016 here (Fig. 7.11). Shimokawa and Kinugasa (1999) conducted a trench excavation at Loc. 025 and identified a nearly vertical fault cutting alluvial sediments. They revealed that the latest faulting event in this section occurred sometime during 1300–1500 yBP.

From Loc. 027 to Loc. 033, the surface rupture was recognized along the preexisting fault scarp that bounded a terrace surface to the west and an alluvial plain to the east (Fig. 7.13). The amount of right-lateral offset gradually decreased from 55 cm at Loc. 030 (Fig. 7.12) to ~30 cm at Locs. 034 and 035 (Fig. 7.14). To the northeast from Loc. 035, the surface rupture became indistinct and died out. There was a branching surface rupture from Loc. 033 toward the east for 440 m with 10–20-cm right-lateral offsets at Locs. 037a, 037b, and 038 (Figs. 7.15, 7.16 and 7.17). We also observed two minor and short surface ruptures subparallel to and 200–400 m southeast of the main rupture. One was a ~150-m-long trace with a right-lateral offset of 8–12 cm and no vertical displacement (Locs. 036 and 037), and the other was a ~10-m-long trace at Loc. 040 with a right-lateral offset of 4 cm and slightly up-to-the-north vertical displacement.

Fig. 7.15 Right-lateral offset (15 cm) of a linear irrigation drain along an edge of the road at Loc. 037a. White arrows indicate the drain. The scale is 40 cm long and oriented parallel to the inferred trend of the surface rupture, but the deformation was so broad that it was difficult to determine the exact location of the rupture

Fig. 7.16 Right-lateral offset (20 cm) of a curb at Loc. 037b. White and red arrows indicate the curb's edge and fault trace, respectively. The scale is 80 cm long and oriented parallel to the inferred trend of the surface rupture. There was no crack on the sidewalk due to repair after the earthquake

Fig. 7.17 Offset on the vegetable field at Loc. 038. Looking to the northwest. White and red arrows indicate a narrow tractor rut as an offset reference and fault trace, respectively. The amount of right-lateral offset was ~20 cm, and the vertical offset was ~10 cm up-on-the-south

Acknowledgements This study was supported by the JSPS grants 18H03601.

References

Ikeda Y, Chida N, Nakata T, Kaneda H, Tajikara M, Takazawa S (2001) 1:25,000 Active fault map in urban area, Kumamoto. Geospatial Information Authority of Japan

Lin A, Chen P, Satsukawa T, Sado K, Takahashi N, Hirata S (2017) Millennium recurrence interval of morphogenic earthquakes on the seismogenic fault zone that triggered the 2016 Mw 7.1 Kumamoto Earthquake, Southwest Japan. Bull Seismol Soc Am 107(6):2687–2702

Shirahama Y, Miyashita Y, Kametaka M, Suzuki Y, Miyairi Y, Yokoyama Y (2020) Detailed paleoseismic history of the Hinagu fault zone revealed by the high-density radiocarbon dating and trenching survey across a surface rupture of the 2016 Kumamoto earthquake, Kyushu, Japan. Island arc 30. https://doi.org/10.1111/iar.12376

Shimokawa K, Kinugasa Y (1999) Paleoseismicity and activity study of the Hinagu fault system. Interim report on active fault and paleoearthquake researches in the 1998 fiscal year, Geological Survey of Japan, no.EQ/99/3, 253–262

Surface Ruptures Along the Kita-Amagi Fault Zone

8

Yasuhiro Kumahara, Daisuke Ishimura, Hiroyuki Tsutsumi, and Nobuhiko Sugito

Abstract

The surface ruptures of the 2016 Kumamoto earthquake from Oike to Itera followed the previously-mapped active fault traces of the Kita-Amagi fault zone. The fault zone cut the Kita-Amagi upland and formed an N20° E-trending depression and a series of isolated uplands that marked the depression's northern edge. In this surface rupture section, both right-lateral and vertical displacements were less than 30 cm, smaller than those of other sections of the 2016 surface rupture. The surface ruptures were located at the base of the uplands, and the sense of vertical displacement at all locations was consistent with the tectonic landform. This observation suggested that the tectonic landform has been formed by repeated coseismic deformation similar to the one during the 2016 earthquake.

Keywords

Surface rupture • 2016 Kumamoto earthquake • Kita-Amagi fault zone • Depression

Supplementary Information
The online version contains supplementary material available at (https://doi.org/10.1007/978-981-19-1150-7_8).

Note
The supplementary material of this book contains geographical information system (GIS) and spreadsheet files of the surface ruptures associated with the mainshock of the 2016 Kumamoto earthquake sequence

Y. Kumahara (✉)
Graduate School of Humanities and Social Sciences, Hiroshima University, 1-1-1, Kagamiyama, Higashi-Hiroshima, Hiroshima, 739-8524, Japan
e-mail: kumakuma@hiroshima-u.ac.jp

D. Ishimura
Department of Geography, Tokyo Metropolitan University, 1-1, Minami-Osawa, Hachioji, Tokyo, 192-0397, Japan
e-mail: ishimura@tmu.ac.jp

H. Tsutsumi
Department of Environmental Systems Science, Faculty of Science and Engineering, Doshisha University, 1-3, Tatara-Miyakodani, Kyotanabe, Kyoto, 610-0394, Japan
e-mail: htsutsum@mail.doshisha.ac.jp

N. Sugito
Faculty of Sustainability Studies, Hosei University, 2-17-1, Fujimi, Chiyoda-ku, Tokyo, 102-8160, Japan
e-mail: nsugito@hosei.ac.jp

North of the surface ruptures in the Shirahata-Oike section, surface ruptures appeared along and near the Kita-Amagi fault zone (Fig. 8.1). Many studies identified the Kita-Amagi fault zone as active (Watanabe et al. 1979; The Research Group for Active Faults of Japan 1980, 1991; Research Group for Active Tectonics in Kyushu 1989; Ikeda et al. 2001; Nakata and Imaizumi 2002). The fault zone cuts the Kita-Amagi upland and forms an N20°E-trending and ~300-m-wide depression as well as a series of isolated uplands that marks the depression's northern edge (Figs. 8.2 and 8.3). The uplands are composed of andesitic lava flow deposits that erupted during 150–260 ka from the Akai Volcano ~5 km to the northeast (see Fig. 9.1 in Chap. 9), which are overlain by fluvial deposits (Ishizaka et al. 1992).

The surface ruptures along the Kita-Amagi fault zone were grouped into four subsections: (1) ruptures along the southern margin of the depression that extended from Shimada to Itera for 1.4 km, (2) ruptures along the northern margin of the depression that extended from Kushijima to Itera for 1.4 km, (3) ruptures along the northern margin of the isolated upland at Higashi-Muta that extended for 1.2 km, and (4) ruptures along the southern margin of the depression at Oike that extended for 0.4 km. These ruptures were located at the base of the uplands, and the sense of vertical displacement at all locations was consistent with the tectonic landform (Figs. 8.4 and 8.5). This observation

Y. Kumahara et al. (eds.), *Surface Ruptures Associated with the 2016 Kumamoto Earthquake Sequence in Southwest Japan*, Advances in Geological Science, https://doi.org/10.1007/978-981-19-1150-7_8

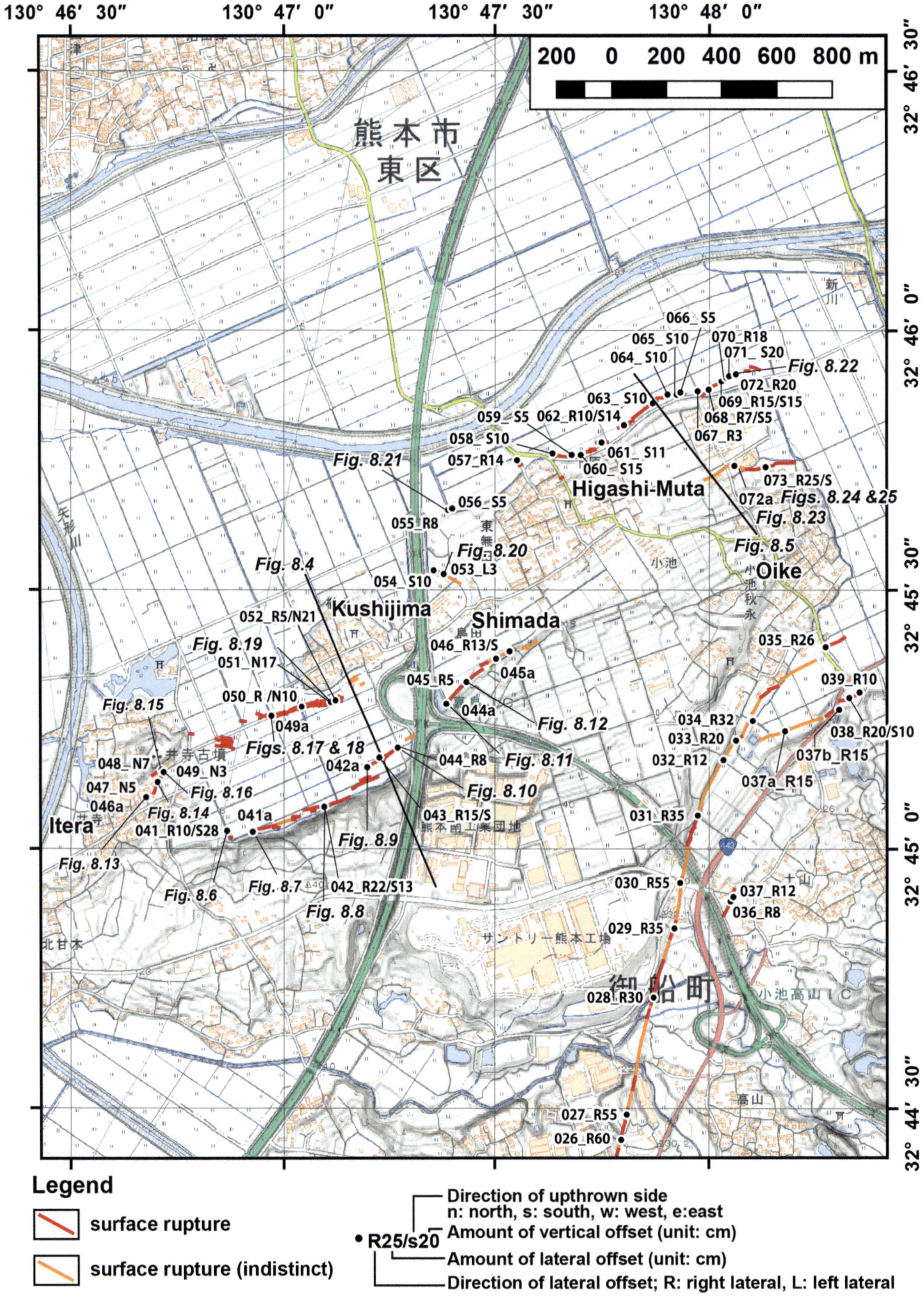

Fig. 8.1 Detailed surface rupture map along the Kita-Amagi fault zone. Strike-slip and vertical displacements at each locality are shown. The background is a half-transparent topographic map of Geospatial Information Authority of Japan (GSI) underlain by a slope map produced from GSI 5-m grid digital elevation models (DEMs)

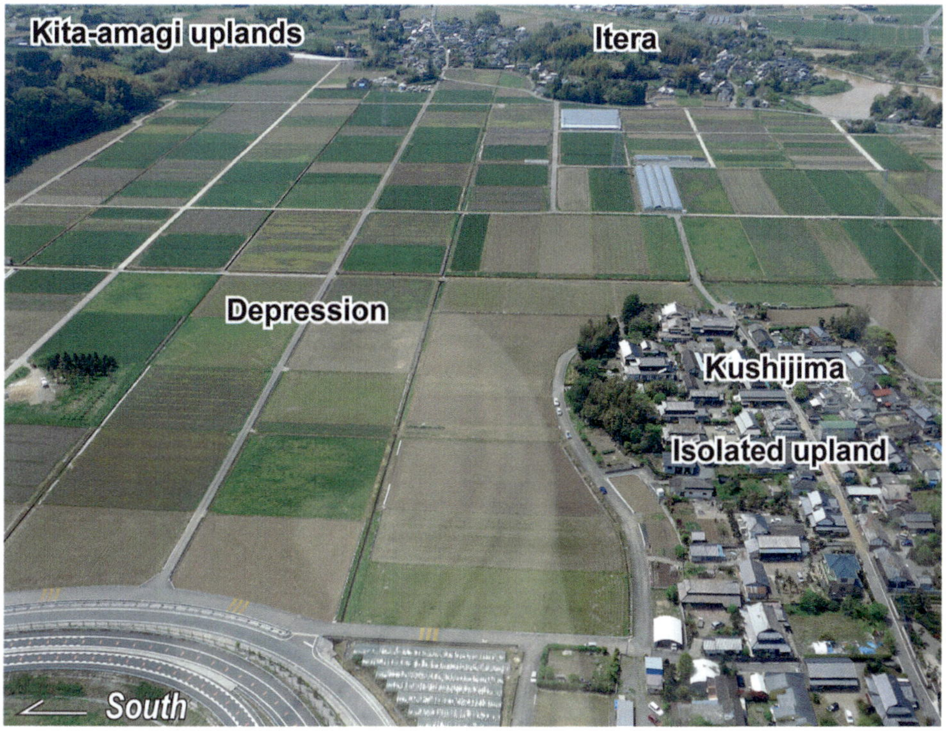

Fig. 8.2 Oblique aerial photograph of the depression and isolated uplands along the Kita-Amagi fault zone taken from a helicopter. Looking to the west

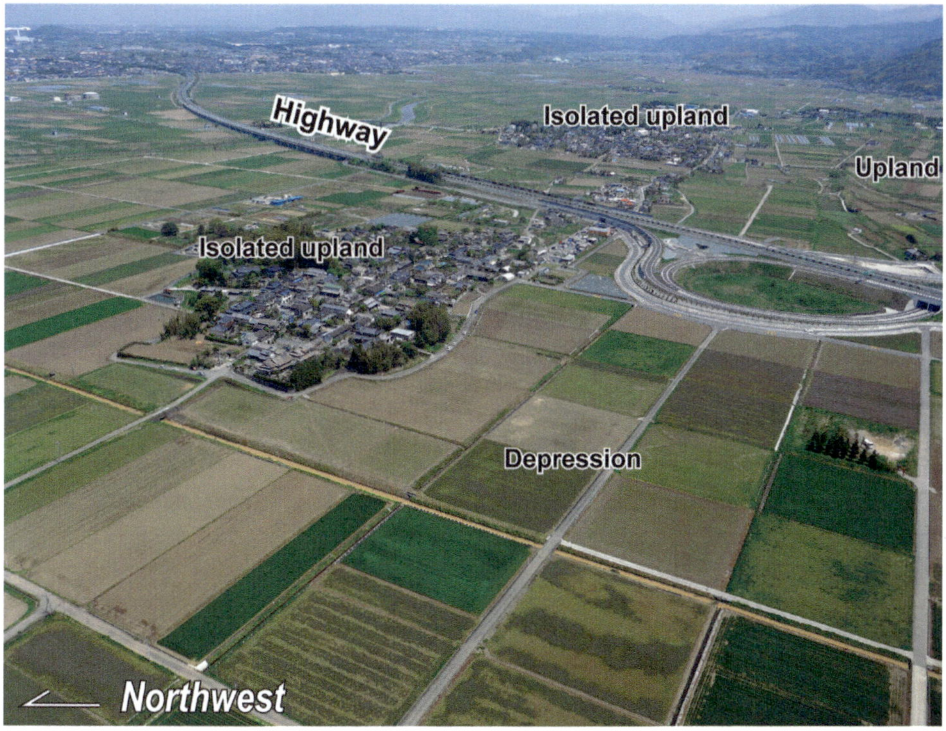

Fig. 8.3 Oblique aerial photograph of the depression along the Kita-Amagi fault zone taken from a helicopter. Looking to the northeast

suggested that the tectonic landform has been formed by repeated coseismic deformation similar to the one during the 2016 earthquake.

The surface ruptures along the southern margin of the depression (Locs. 041 to 042) were characterized by a relatively simple trace with a left-stepping *en echelon* structure (Figs. 8.1, 8.6, 8.7 and 8.8). We observed consistent right-lateral offset along this section, 10 cm at Loc. 041 (Fig. 8.6) and 22 cm at Loc. 042 (Fig. 8.8). Vertical displacement was up on the south, 28 cm at Loc. 041 and 13 cm at Loc. 042. To the east, we mapped several rupture traces within a ~50-m-wide zone with 8 to 15 cm right-lateral offset (Figs. 8.9 and 8.10). Further east, there was a faint single trace that cut across a road at Loc. 044a

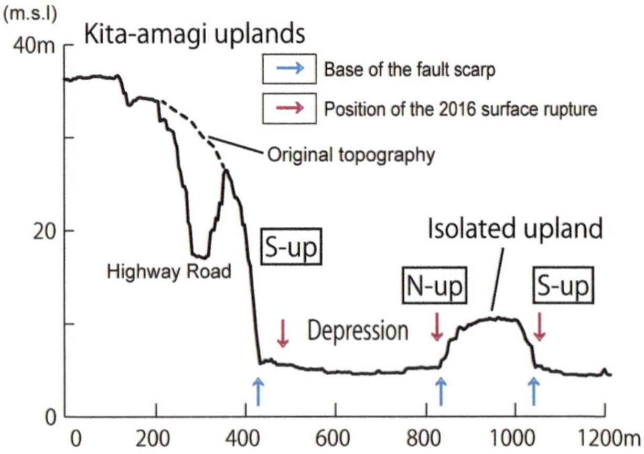

Fig. 8.4 Topographic profile across the Kita-Amagi uplands and the tectonic depression around Kushijima based on 5-m grid DEMs of GSI. Blue and red arrows indicate the base of the fault scarp and the position of the 2016 surface rupture, respectively. See Fig. 8.1 for the location and extent of the profile

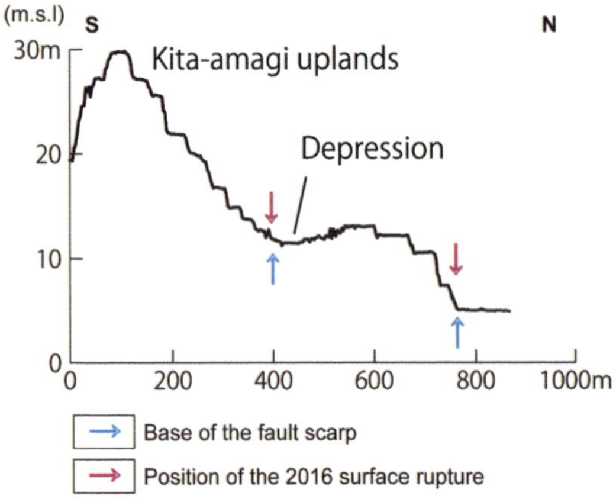

Fig. 8.5 Topographic profile across the fault scarps at the base of the Kita-Amagi uplands and the tectonic depression around Oike based on 5-m grid DEMs of GSI. Blue and red arrows indicate the base of the fault scarp and the position of the 2016 surface rupture, respectively. See Fig. 8.1 for the location and extent of the profile

Fig. 8.6 Photograph of the offset white line on the paved road at Loc. 041. We measured 10 cm right-lateral offset with up-on-the-south vertical displacement of 28 cm. White and red arrows indicate the reference line and fault trace, respectively

Fig. 8.7 Photograph of the surface rupture on the paved road at Loc. 041a. Red arrows denote the fault trace

Fig. 8.8 Photograph of right-lateral offset (22 cm) with up-on-the-south vertical displacement (13 cm) of a gutter at Loc. 042. White and red arrows denote the offset gutter and fault trace, respectively

Fig. 8.9 Photograph of a ~50-m-wide zone of distributed rupture traces at Loc. 042a

Fig. 8.10 Photograph of the surface rupture at Loc. 044. Red arrows indicates the fault trace

Fig. 8.11 Photograph of right-lateral offset with an up-on-the-south displacement of the curb at Loc. 044a. White and red arrows indicate the offset curb and fault trace, respectively

Fig. 8.12 Photograph of right-lateral offset (5 cm) at Loc. 045. White and red arrows indicate the offset marker and fault trace, respectively

(Fig. 8.11) and the fault extended with a left-stepping *en echelon* structure with 5 cm right-lateral offset at Loc. 045 (Fig. 8.12) to 13 cm at Loc. 046. At Loc. 045a, where the trace of surface rupture was ambiguous, Maruyama et al. (2019) dug a trench and identified distinct vertical deformation larger than that of the 2016 earthquake sequence.

The surface rupture along the northern margin of the depression extended from Loc. 046a (Fig. 8.13), where the trace was recognized as a small crack. The trace from Loc. 046a to Loc. 049 was identified as open cracks with north-side-up vertical displacement less than 10 cm (Figs. 8.14, 8.15 and 8.16). Several sub-parallel cracks within a ~15-m-wide zone appeared on the alluvial plain

Fig. 8.13 Photograph of the western termination of the surface rupture along the northern margin of the depression at Loc. 046a. Red arrows indicate the surface rupture

Fig. 8.14 Photograph of the open cracks at Loc. 047. Red arrows indicate the fault trace

Fig. 8.15 Photograph of north-side-up vertical displacement (~7 cm) on the paved road at Loc. 048. Red arrows indicate the surface rupture

Fig. 8.16 Photograph of north-side-up vertical displacement (~3 cm) on the paved road at Loc. 049. Red arrows indicate the surface rupture. Loc. 048 is ~12 m to the north of Loc. 049

Fig. 8.17 Photograph of the open cracks west of Loc. 049a. Red arrows indicate the ruptures

Fig. 8.18 Photograph of the open cracks east of Loc. 049a

between the isolated uplands from Loc. 049a to Loc. 052 for 500 m (Figs. 8.17, 8.18 and 8.19). We identified 10 to 21 cm north-side-up vertical displacement at these sites.

There was an N60°W-striking, ~130-m-long surface rupture in the lowland between the uplands at Kushijima and Higashi-Muta. The short trace trended conjugate to the axis of the uplands. At Loc. 053, we found a left-lateral offset of ~3 cm of a footpath (Fig. 8.20).

The surface rupture along the northern edge of the isolated upland at Higashi-Muta extended for ~1.2 km from Loc. 055 to Loc. 072 and consisted of several open cracks

Fig. 8.19 Photograph of the open cracks at Loc. 051

Fig. 8.20 Photograph of left-lateral offset (3 cm) of a footpath at Loc. 053. White and red arrows indicate the edge of the footpath and surface rupture, respectively

Fig. 8.21 Photograph of right-lateral offset (8 cm) of an edge of the paved road at Loc. 055. White and red arrows indicate the edge of the paved road and surface rupture, respectively

sub-parallel to the strike of the uplands with a left-stepping *en echelon* structure. South-side-up, 5 to 20 cm vertical displacement was commonly observed along the trace. The right-lateral slip was 3 to 20 cm between Loc. 055 (Fig. 8.21) and Loc. 072 (Fig. 8.22).

The ~0.4-km-long rupture trace along the southern margin of the depression at Oike showed a right-lateral offset with minor south-side-up vertical displacement (Fig. 8.23). At Loc. 073, we observed open cracks with a left-stepping *en echelon* structure (Fig. 8.24) and measured ~25 cm right-lateral offset (Fig. 8.25).

Acknowledgements This study was supported by the JSPS grants 18H03601.

Fig. 8.22 Photograph of right-lateral offset (20 cm) with up-on-the-south vertical displacement on the footpath at Loc. 072. White and red arrows indicate the footpath and surface rupture, respectively

Fig. 8.23 Photograph of the surface rupture at Loc. 072a. South-side-up vertical displacement is observed. Red arrows indicates the fault trace

Fig. 8.24 Photograph of the open cracks with a left-stepping *en echelon* structure at Loc. 073

Fig. 8.25 Photograph of right-lateral offset (25 cm) of aligned posts of the grape shelf at Loc. 073. White and red arrows indicate the posts and surface rupture, respectively

References

Ikeda Y, Chida N, Nakata T, Kaneda H, Tajikara M, Takazawa S (2001) 1:25,000 Active Fault Map in Urban Area, Kumamoto. Geospatial Information Authority of Japan

Ishizaka S, Watanabe K, Takada H (1992) Subsidence rate of the underground Quaternary system for the past 150,000 years in the Kumamoto Plain, Kyushu, Japan. Quat Res (Daiyonki Kenkyu) 31:91–99. https://doi.org/10.4116/jaqua.31.91

Maruyama T, Saito M, Komine Y, Kametaka M (2019) Paleoseismological trenching and detailed mapping of the surface rupture appeared along the Kita-amagi fault during the 2016 Kumamoto earthquake at Shimada, Mashiki Town, Kumamoto Prefecture, southwest Japan. Active Fault Res 50:13–31

Nakata T, Imaizumi T (eds) (2002) Digital active fault map of Japan. University of Tokyo Press, Tokyo

The Research Group for Active Faults of Japan (ed) (1980) Active faults in Japan: sheet maps and inventories. University of Tokyo Press, Tokyo

The Research Group for Active Faults of Japan (ed) (1991) Active faults in Japan: sheet maps and inventories (Revised edition). University of Tokyo Press, Tokyo

The Research Group for Active Tectonics in Kyushu (ed) (1989) Active tectonics in Kyushu. University of Tokyo Press, Tokyo

Watanabe K, Momikura Y, Tsuruya K (1979) Active faults and parasitic eruption centers on the west flank of Aso Caldera, Japan. Quat Res (Daiyonki Kenkyu) 18:89–101. https://doi.org/10.4116/jaqua.18.89

Surface Ruptures Along the Southern Part of the Futagawa Fault

9

Yasuhiro Kumahara, Hideaki Goto, and Hiroyuki Tsutsumi

Abstract

This chapter describes the characteristics of the surface ruptures of the 2016 Kumamoto earthquake along the 6-km-long southern part of the previously mapped Futagawa fault from Shimotogawa to Shimojin, with photographs and detailed maps. In contrast to the continuous surface traces of the geomorphologically evident Futagawa fault, the coseismic surface ruptures exhibited a left-stepping *en echelon* structure with short conjugate left-lateral faults. Right-lateral offset was dominant on the northeast-southwest trending main trace, and the maximum amount of the offset in the section was 90 cm. Up-to-the-southeast vertical displacement was common in the section. On the other hand, short conjugate faults showed left-lateral offset, with the amount of displacement being almost the same as the right-lateral offset along the main fault trace. In the northern part of this section, two sub-parallel surface rupture traces were found: one along the base of the mountain and the other in the middle of the mountain slope.

Keywords

Surface rupture • 2016 Kumamoto earthquake • Futagawa fault

Northeast of the Shirahata-Oike section (Chap. 7) and east of the Kita-Amagi fault zone (Chap. 8), we observed a 6-km-long surface rupture from Shimotogawa to Shimojin along the mountain's northern foot or across the alluvial plains close to the mountain front (Figs. 9.1 and 9.2). The rupture followed roughly the southern part of the Futagawa fault, which was known as an active fault before the earthquake (e.g., The Research Group for Active Faults of Japan 1991). The rupture traces were divided into five left-stepping segments separated by 0.2- to 0.7-km-wide steps; the Shimotogawa, Kamitogawa, Goraku-Minami, Akai-Fukuhara, and Tanaka-Shimojin segments (Figs. 9.1 and 9.2). All the segments except the Shimotogawa segment showed right-lateral strike-slip with N50°E to N70°E strikes. The surface rupture along the Shimotogawa segment slipped mainly up-on-the-south, striking N78°E. There were four other traces as follows. There was a 2 km-long trace from Hirata to Shimojin, sub-parallel to and south of the trace of the Tanaka-Shimojin segment (Fig. 9.2). Two parallel left-lateral strike-slip ruptures trending E to N30°W were observed for 800 m at Fukuhara (Figs. 9.1 and 9.2). An N65°W-trending 600-m-long conjugate rupture appeared from Shimojin to Jicchu (Fig. 9.2). In the following, we describe the detailed characteristics of these surface ruptures from south to north.

The 0.7-km-long and N78°E-trending Shimotogawa segment cut across an alluvial plain and was located at

Supplementary Information
The online version contains supplementary material available at (https://doi.org/10.1007/978-981-19-1150-7_9).

Note
The supplementary material of this book contains geographical information system (GIS) and spreadsheet files of the surface ruptures associated with the mainshock of the 2016 Kumamoto earthquake sequence

Y. Kumahara (✉)
Graduate School of Humanities and Social Sciences, Hiroshima University, 1-1-1, Kagamiyama, Higashi-Hiroshima, Hiroshima, 739-8524, Japan
e-mail: kumakuma@hiroshima-u.ac.jp

H. Goto
Graduate School of Humanities and Social Sciences, Hiroshima University, 1-2-3, Kagamiyama, Higashi-Hiroshima, Hiroshima, 739-8522, Japan
e-mail: hgoto@hiroshima-u.ac.jp

H. Tsutsumi
Department of Environmental Systems Science, Faculty of Science and Engineering, Doshisha University, 1-3, Tatara-Miyakodani, Kyotanabe, Kyoto, 610-0394, Japan
e-mail: htsutsum@mail.doshisha.ac.jp

Y. Kumahara et al. (eds.), *Surface Ruptures Associated with the 2016 Kumamoto Earthquake Sequence in Southwest Japan*, Advances in Geological Science, https://doi.org/10.1007/978-981-19-1150-7_9

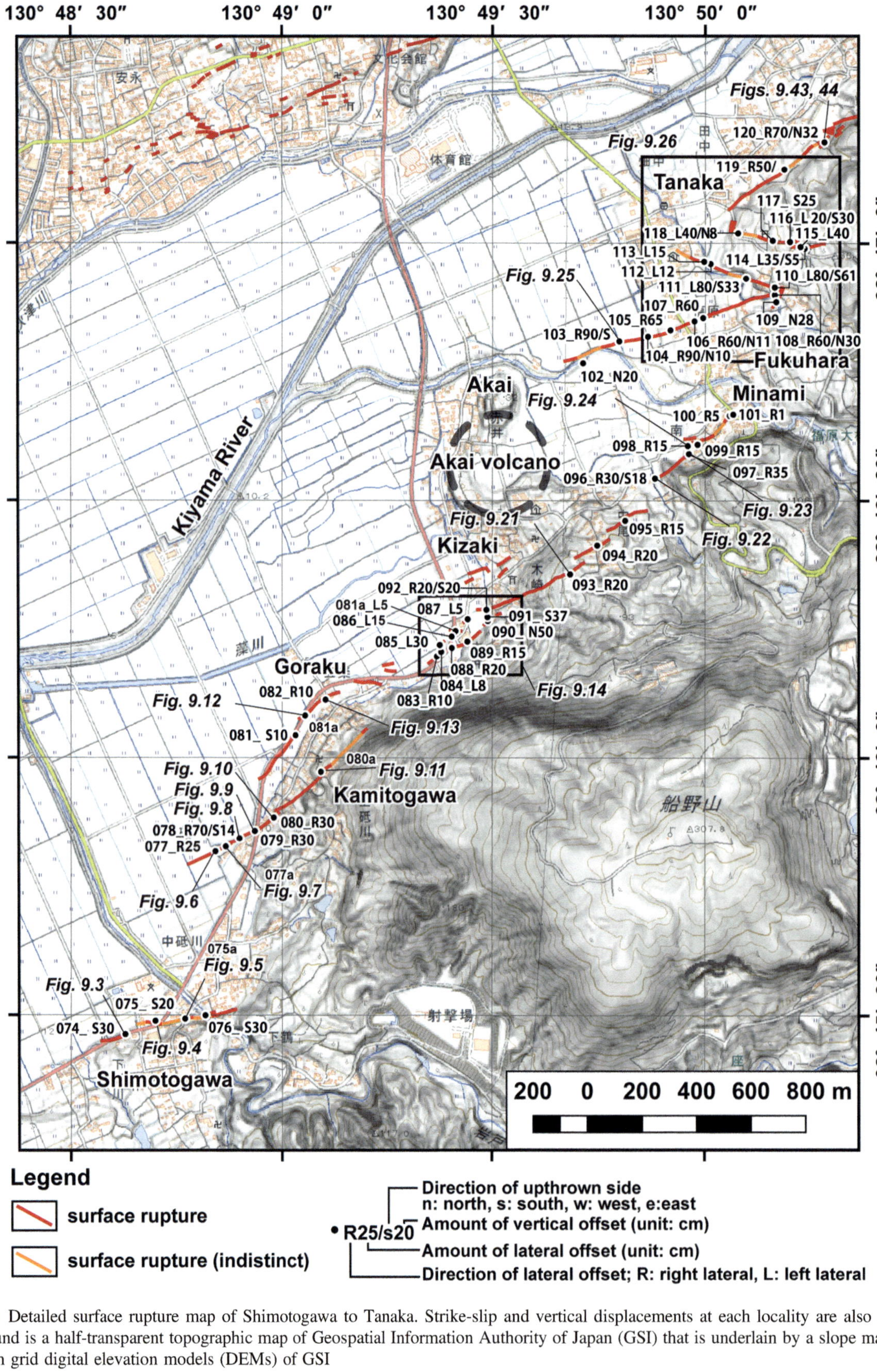

Fig. 9.1 Detailed surface rupture map of Shimotogawa to Tanaka. Strike-slip and vertical displacements at each locality are also shown. The background is a half-transparent topographic map of Geospatial Information Authority of Japan (GSI) that is underlain by a slope map produced from 5-m grid digital elevation models (DEMs) of GSI

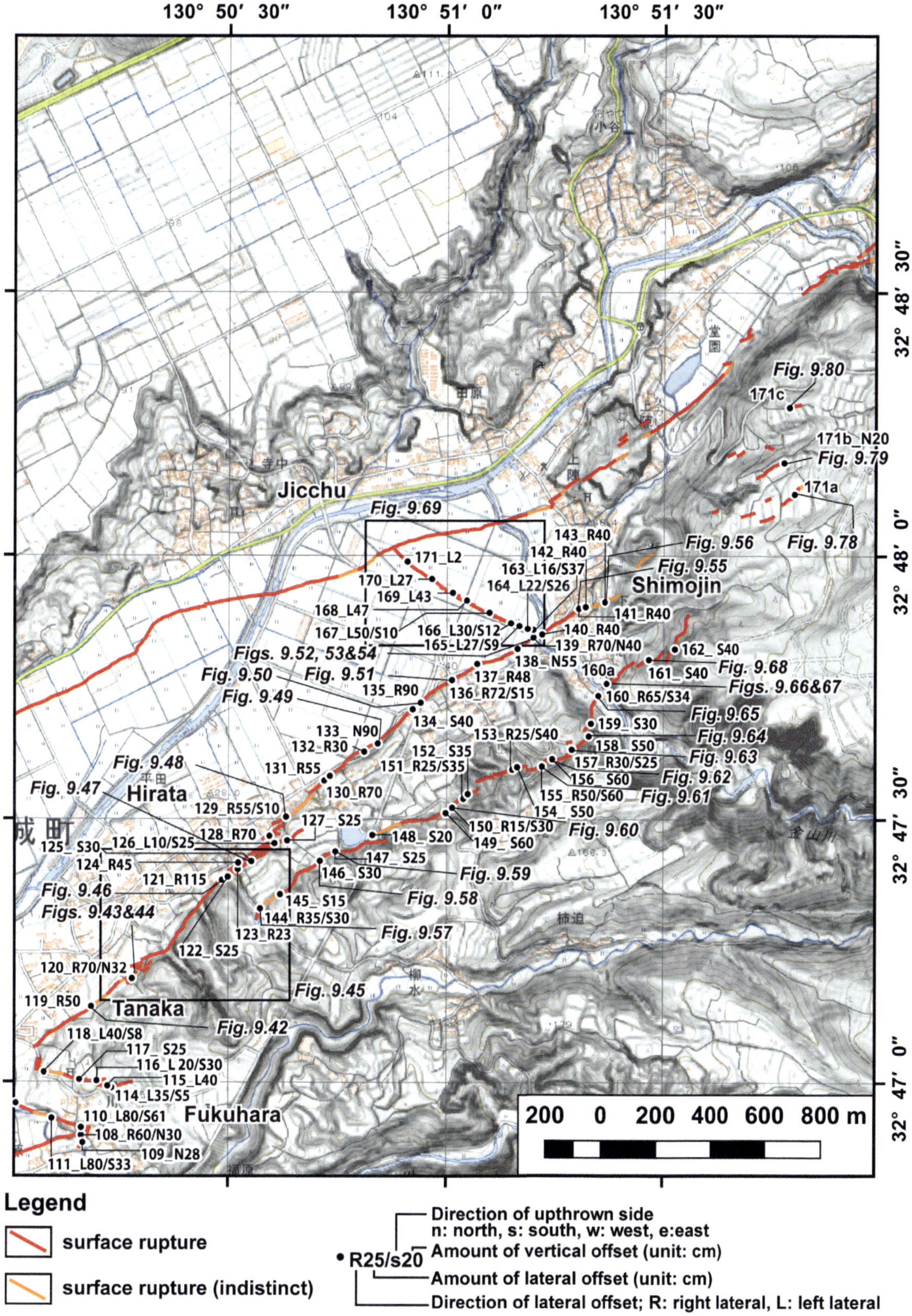

Fig. 9.2 Detailed surface rupture map from Fukuhara and Shimojin. Strike-slip and vertical displacements at each locality are also shown. The base map is the same as in Fig. 9.2

0.5 km from the northern end of the Shirahata-Oike section (see Chap. 7 for detail). The displacement along the segment was 20–30 cm up-on-the-south at Locs. 074, 075, and 076 (Figs. 9.3, 9.4, and 9.5). Figure 9.4 is a photo taken at Loc. 075 just after the mainshock and shows up-on-the-south displacement and left-lateral strike-slip based on the offset of white lines on the road. Here, the trace struck E–W, which was slightly rotated from the neighboring traces (N78°E), suggesting that the change in strike caused the localized lateral offset.

The N58°E-trending Kamitogawa segment was ~1 km long and was located ~500 m north of the Shimotogawa segment. Between Locs. 077 and 080, the right-lateral displacement was 25–30 cm (Figs. 9.6, 9.7, 9.9, and 9.10) except for Loc. 078 (Fig. 9.8). At Loc. 078, we observed the maximum right-lateral offset of 70 cm on this segment with 14-cm up-on-the-south displacement on the cultivated land. The trace between Locs. 078 and 079 followed the base of the preexisting 3-m-high fault scarp. Northeast from Loc. 080, the trace appeared as open cracks along the base of the mountain slope (Fig. 9.11).

The 2.35-km-long and N50°E-trending Goraku-Minami segment consisted of a series of left-stepping ruptures. We observed a sand blow on the paddy field at Loc. 081a

Fig. 9.3 Surface rupture (red arrows) on the pavement at Loc. 074 along the Shimotogawa segment. South-side-up vertical displacement (30 cm) was observed

Fig. 9.4 Surface rupture on the paved road at Loc. 075 along the Shimotogawa segment. South-side-up vertical displacement with left-lateral slip was observed. The white and red arrows indicate offset markers and the surface ruptures, respectively

Fig. 9.5 Surface rupture (red arrows) at Loc. 075a along the Shimotogawa segment. South-side-up vertical displacement (20 cm) was observed

Fig. 9.6 Right-lateral offset of the edge of the paved road at Loc. 077 along the Shimotogawa segment. The amount of right-lateral offset was ~25 cm. The white and red arrows indicate an offset marker and surface ruptures, respectively

Fig. 9.7 Left-stepping open cracks (red arrows) at Loc. 077a along the Shimotogawa segment. The ruptures extended under the house

Fig. 9.8 Right-lateral offset (70 cm) with up-on-the-south (14 cm) displacement on the cultivated land at Loc. 078 along the Shimotogawa segment. The white and red arrows indicate an offset marker and the surface ruptures, respectively

Fig. 9.9 Left-stepping ruptures (red arrows) at the base of the existed fault scarp at Loc. 079 along the Shimotogawa segment

Fig. 9.10 Right-lateral offset (30 cm) of the concrete wall at Loc. 080 along the Shimotogawa segment. The white and red arrows indicate an offset marker and the surface rupture, respectively

Fig. 9.11 Surface rupture (red arrows) in the chestnut orchard at Loc. 080a along the Shimotogawa segment

(Fig. 9.12). At Goraku, the surface rupture was indistinct with less than 10-cm right-lateral or up-on-the-south displacement (Fig. 9.13). A complex rupture geometry was observed at Kizaki, such as a minor pressure ridge between the main traces and minor conjugate left-lateral ruptures (Fig. 9.14). Several conjugate traces trended N80°W to E–W north of the primary traces, with 5- to 30-cm left-lateral slip at Locs. 085, 086, 086a, and 087 (Figs. 9.15 and 9.16). The

Fig. 9.12 Sand blow on the paddy field at Loc. 081a along the Goraku-Minami segment

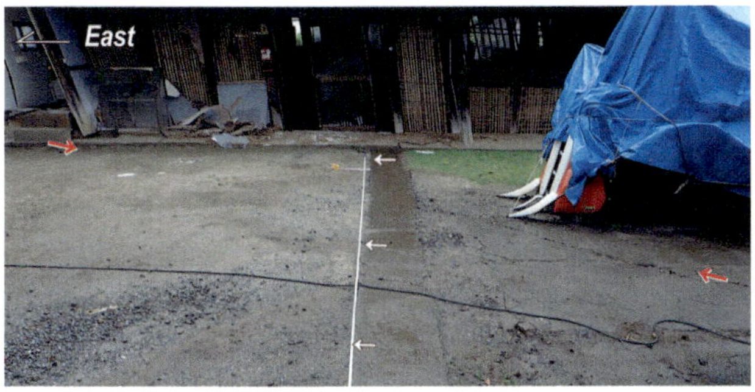

Fig. 9.13 Right-lateral offset (10 cm) of concrete pavement at Loc. 082 along the Goraku-Minami segment. The white and red arrows indicate an offset marker and the surface rupture, respectively

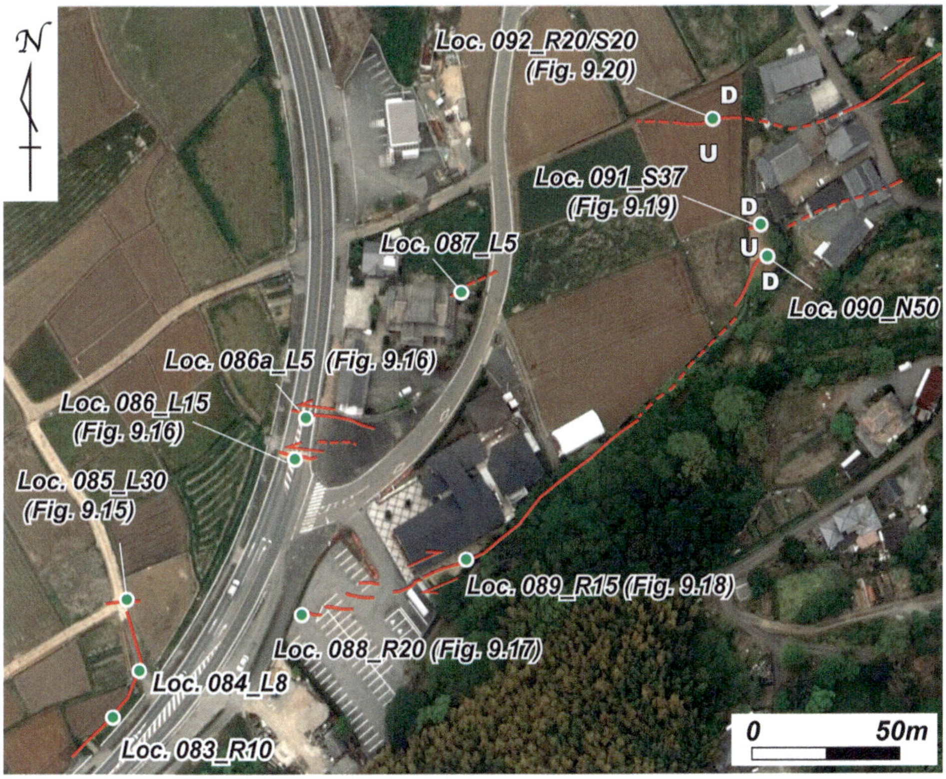

Fig. 9.14 Map of the detailed rupture traces at Kizaki along the Goraku-Minami segment. The solid red and broken red lines are confirmed and site-indistinct traces, respectively. The labels U and D represent the uplifted and the downthrown sides, respectively. The background air photo was taken by the Geospatial Information Authority of Japan. The location of the map is shown in Fig. 9.1

Fig. 9.15 Left-lateral offset (30 cm) of the concrete wall at Loc. 085 along the Goraku-Minami segment. The white and red arrows indicate an offset marker and the surface rupture, respectively

Fig. 9.16 Left-lateral offset (20 cm) of the road's centerline at Locs. 086 and 086a along the Goraku-Minami segment. The white and red arrows indicate an offset marker and the surface ruptures, respectively. Locs. 085 and 088 are also visible in the background

Fig. 9.17 Left-stepping surface ruptures (open cracks; red arrows) on the asphalt paved parking area at Loc. 088 along the Goraku-Minami segment

N50°E-trending main trace followed the base of the hill with 10- to 20-cm right-lateral slip at Locs. 083, 088, 089, and 092 (Figs. 9.17 and 9.18). At Loc. 088, we identified left-stepping open cracks on the paved parking lot. We recognized an uplifted zone with 37- to 50-cm vertical displacement between two minor parallel traces at Locs. 090 and 091 (Fig. 9.19). The primary trace disappeared within ~50 m from Loc. 091, and another trace appeared ~25 m to the north. At Loc. 092, we measured 20-cm right-lateral offset and 20-cm up-on-the-south vertical displacement based on tractor ruts (Fig. 9.20). Several left-stepping traces cut across ridges from Kizaki to Minami.

Fig. 9.18 Left-stepping surface ruptures (open cracks; red arrows) at Loc. 089 along the Goraku-Minami segment

Fig. 9.19 Uplifted vegetable field between two parallel traces (red arrows) at Locs. 090 and 091 along the Goraku-Minami segment

Fig. 9.20 Right-lateral offset (20 cm) of tractor ruts and south-up vertical separation at Loc. 093 along the Goraku-Minami segment. The white and red arrows indicate an offset marker and the surface rupture, respectively

The amount of right-lateral offset was 20–35 cm (Figs. 9.21, 9.22 and 9.23), but decreased from Loc. 098 (Fig. 9.24) toward its northeastern end at Minami.

About 400 m north of the Goraku-Minami segment, the Akai-Fukuhara segment trended N20°E and extended for ~750 m across an alluvial plain. The displacement along

Fig. 9.21 Right-lateral offset (30 cm) of the aligned stepping grass patches in the persimmon orchard at Loc. 093 along the Goraku-Minami segment. The white and red arrows indicate an offset marker and the surface ruptures, respectively

Fig. 9.22 Right-lateral offset (30 cm) of a plow boundary and south-up vertical separation (18 cm) at Loc. 096 along the Goraku-Minami segment. The white and red arrows indicate an offset marker and the surface rupture, respectively

Fig. 9.23 Right-lateral offset (35 cm) of gutter and road at Loc. 097 along the Goraku-Minami segment. The white and red arrows indicate an offset marker and the surface rupture, respectively

Fig. 9.24 Right-lateral offset (15 cm) of gutter and road at Loc. 098 along the Goraku-Minami segment. The white and red arrows indicate an offset marker and the surface rupture, respectively

Fig. 9.25 Right-lateral offset (90 cm) of the paddy field boundary at Loc. 103 along the Akai-Fukuhara segment. The white and red arrows indicate an offset marker and the surface rupture, respectively

the trace was 60–90 cm right-laterally (Figs. 9.25, 9.26, 9.27, 9.28, 9.29, 9.30, 9.31, and 9.32). The vertical displacement along the trace was south-up on the western half and north-up on the eastern half, consistent with the expected pattern of vertical displacement for a right-lateral strike slip fault. At the Akai-Fukuhara segment's eastern termination, we found an N17°W-trending conjugate left-lateral strike-slip rupture. At Locs. 108 and 110

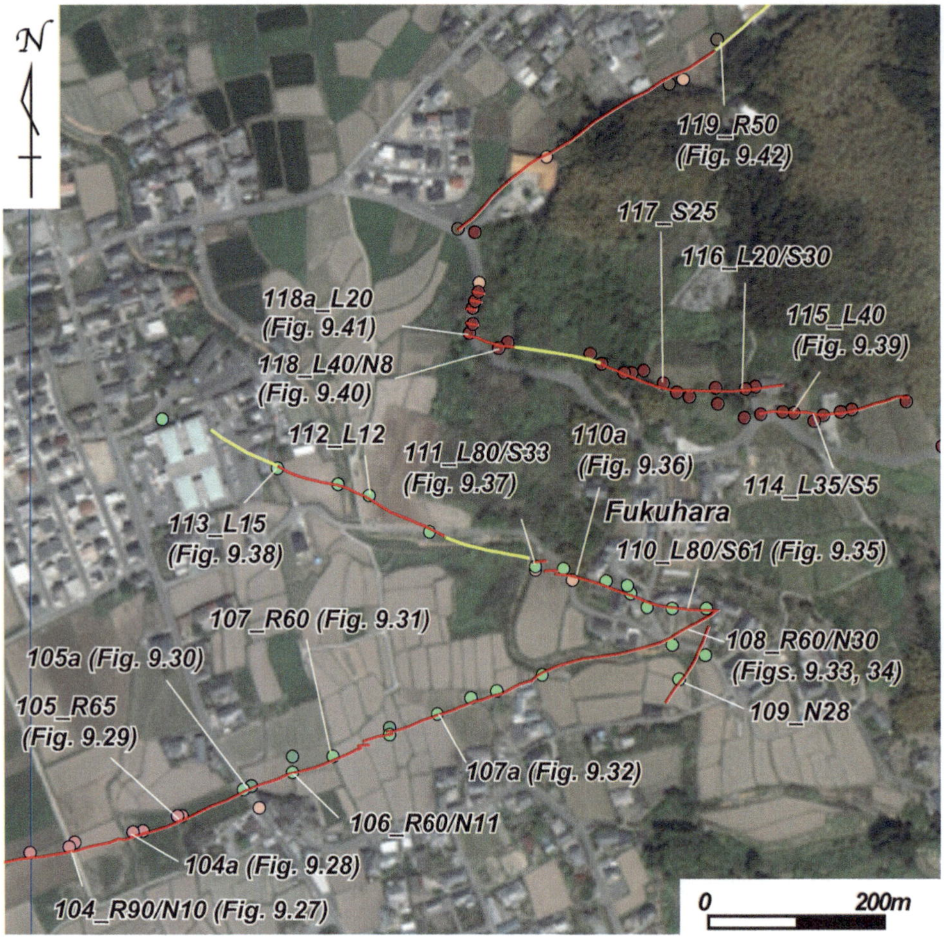

Fig. 9.26 Map of the detailed surface rupture traces around Fukuhara along the Akai-Fukuhara segment. Red and yellow lines are confirmed and site-indistinct traces, respectively. The small circles indicate the measuring points by hand-held GNSS devices. The different colors correspond to the different observers. The location of the map is shown in Fig. 9.1. The background air photo was taken by the Geospatial Information Authority of Japan

Fig. 9.27 Right-lateral offset (90 cm) of gutter and road with south-up displacement (10 cm) at Loc. 104 along the Akai-Fukuhara segment. The white and red arrows indicate an offset marker and the surface rupture, respectively. The location of the photo is shown in Fig. 9.26

(Figs. 9.33, 9.34, and 9.35) in the private garden, we identified a junction of the up-on-the-north right-lateral rupture of the Akai-Fukuhara segment and up-on-the-south left-lateral rupture. The area between the two ruptures was uplifted. The conjugate left-lateral rupture was ~500 m long with the maximum left-lateral slip of 80 cm (Figs. 9.36,

Fig. 9.28 Photo of the surface rupture taken from Loc. 104a along the Akai-Fukuhara segment. The location of the photo is shown in Fig. 9.26

Fig. 9.29 Right-lateral offset (65 cm) of the footpath in the paddy field at Loc. 105 along the Akai-Fukuhara segment. The white and red arrows indicate an offset marker and the surface rupture, respectively. The location of the photo is shown in Fig. 9.26

Fig. 9.30 Right-lateral offset of a white line on the road at Loc. 105a along the Akai-Fukuhara segment. The white and red arrows indicate an offset marker and the surface rupture, respectively. The location of the photo is shown in Fig. 9.26

9.37, and 9.38). In addition, there was another 400-m-long rupture (called the northern rupture) 200 m north of and parallel to the rupture (Figs. 9.39, 9.40 and 9.41). The offset on the northern rupture was left-lateral with up-on-the-south vertical separation with the maximum left-lateral offset of 40 cm. The western end of the northern rupture is

Fig. 9.31 Right-lateral offset (60 cm) of the boundary of paddy field at Loc. 107 along the Akai-Fukuhara segment. The red arrows indicate surface ruptures. The location of the photo is shown in Fig. 9.26

Fig. 9.32 Right-lateral offset of the concrete-made boundary of paddy field at Loc. 107a along the Akai-Fukuhara segment. The white and red arrows indicate an offset marker and the surface ruptures, respectively. The location of the photo is shown in Fig. 9.26

Fig. 9.33 Junction of right-lateral rupture and conjugated left-lateral rupture in the private garden at Locs. 108 and 110 at Fukuhara. The red arrows indicate surface ruptures. The location of the photo is shown in Fig.9.26

located immediately south of the southern end of the Tanaka-Shimojin segment.

The 3-km-long Tanaka-Shimojin segment trended N40°E along the northern foot of the mountain. The right-lateral offset along the trace was 30–90 cm with minor north- or south-side-up vertical separations (Figs. 9.42, 9.43, 9.44, 9.45, 9.46, 9.47, 9.48, 9.49, 9.50, 9.51, 9.52, 9.53, 9.54, 9.55, and 9.56) The amount of right-lateral displacement

Fig. 9.34 Right-lateral offset (60 cm) of the ditch and north-up vertical separation (30 cm) at Loc. 108 along the Akai-Fukuhara segment. The white and red arrows indicate an offset marker and the surface rupture, respectively. The location of the photo is shown in Fig. 9.26

Fig. 9.35 Left-lateral offset (80 cm) of the ditch and south-up vertical separation (61 cm) at Loc. 110. The white and red arrows indicate an offset marker and surface ruptures, respectively. The location of the photo is shown in Fig. 9.26

Fig. 9.36 Left-lateral rupture (red arrows) with a right-stepping structure on the road and paddy field at Loc. 110a. The location of the photo is shown in Fig. 9.26

Fig. 9.37 Left-lateral offset (80 cm) of the rice paddy dike and south-up vertical separation (33 cm) at Loc. 111. The white and red arrows indicate an offset marker and the surface rupture, respectively. The location of the photo is shown in Fig. 9.26

Fig. 9.38 Left-lateral offset (15 cm) of the edge of paddy field at Loc. 113. The white and red arrows indicate an offset marker and the surface ruptures, respectively. The location of the photo is shown in Fig. 9.26

Fig. 9.39 Left-lateral offset (40 cm) of the footpath at Loc. 115. The white and red arrows indicate an offset marker and the surface rupture, respectively. The location of the photo is shown in Fig. 9.26

Fig. 9.40 Left-lateral offset (40 cm) of the furrow and north-up vertical separation (8 cm) at Loc. 118. The white and red arrows indicate an offset marker and the surface rupture, respectively. The location of the photo is shown in Fig. 9.26

Fig. 9.41 Left-lateral offset (20 cm) of the white line on the road at Loc. 118a. The white and red arrows indicate an offset marker and the surface rupture, respectively. The location of the photo is shown in Fig. 9.26

Fig. 9.42 Right-lateral offset (50 cm) of the footpath at Loc. 119 along the Tanaka-Shimojin segment. The white and red arrows indicate an offset marker and the surface rupture, respectively. Yoshioka et al. (2007) conducted a trench survey here before the 2016 earthquake. The location of the photo is shown in Fig. 9.26

Fig. 9.43 Right-lateral offset (70 cm) of irrigation channel and north-up vertical separation (32 cm) at Loc. 120 along the Tanaka-Shimojin segment. The white and red arrows indicate an offset marker and the surface ruptures, respectively

Fig. 9.44 Surface ruptures on the playground at Loc. 120 along the Tanaka-Shimojin segment, where the hill had been artificially cut and made flat

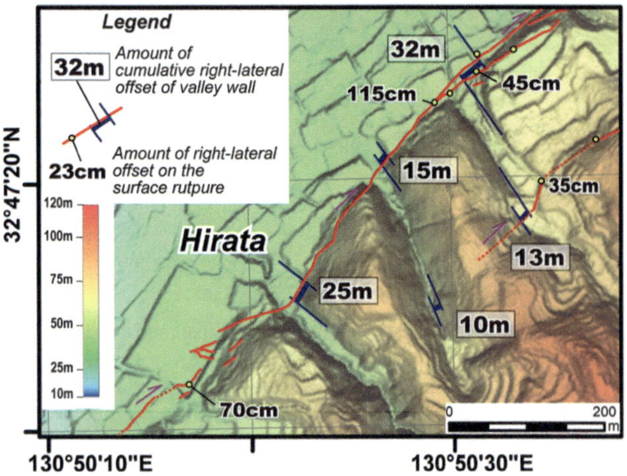

Fig. 9.45 Detailed map of the surface ruptures, amount of coseismic right-lateral offset, and amount of cumulative right-lateral offset of stream channels around Hirata. The location of the map is shown in Fig. 9.2

Fig. 9.46 Right-lateral offset (115 cm) of the fence at Loc. 121 along the Tanaka-Shimojin segment. The white and red arrows indicate an offset marker and the surface rupture, respectively

Fig. 9.47 Rupture (red arrows) on the hill slope with south-up displacement (30 cm) at Loc.125 along the Tanaka-Shimojin segment

increased toward the north. Three stream channels at Hirata are offset right-laterally, ranging from 15 to 32 m, indicating repeated fault movement (Fig. 9.45). Yoshioka et al. (2007) carried out a trenching study at Loc. 119 (Tanaka site; Fig. 9.42). A near-vertical fault exposed on the trench wall cut gravel and silt layers. Radiocarbon dates suggested that the last faulting event before the 2016 event occurred sometime between 12,000 and 2200 years ago (Yoshioka et al. 2007).

About 250–600 m southeast of and sub-parallel to the Tanaka-Shimojin segment, we mapped 1.2-km-long another surface rupture trace in the middle of the hill slope from

Fig. 9.48 Near-horizontal striations on the fault plane of the surface rupture at Loc. 129 along the Tanaka-Shimojin segment

Fig. 9.49 Left-stepping surface ruptures (red arrows) with north-up displacement at Loc. 133 along the Tanaka-Shimojin segment. The local subsidence is responsible for the wet brownish area immediately south of the ruptures

Fig. 9.50 Right-lateral offset (90 cm) of the centerline of the road at Loc. 135 along the Tanaka-Shimojin segment. The white and red arrows indicate an offset marker and the surface rupture, respectively

Fig. 9.51 Right-lateral offset (72 cm) of the road's edge at Loc. 136 along the Tanaka-Shimojin segment. The white and red arrows indicate an offset marker and the surface rupture, respectively

Fig. 9.52 Surface ruptures (red arrows) with right-lateral offset on the paddy field at Loc. 139 along the Tanaka-Shimojin segment

Fig. 9.53 Right-lateral offset (70 cm) of the footpath with north-up-thrown separation (40 cm) on the vegetable field at Loc. 139 along the Tanaka-Shimojin segment. The white and red arrows indicate an offset marker and the surface rupture, respectively

Hirata to Shimojin (Figs. 9.57, 9.58, 9.59, 9.60, 9.61, 9.62, 9.63, 9.64, and 9.65). The rupture was characterized by right-lateral normal faulting and followed a geomorphic boundary between the gentle slope in the north and much steeper slope in the south. The right-lateral offset ranged from 35 to 65 cm in the west and increased toward the east to Loc. 160. The south-side-up vertical separation was 25–30 cm in the west, 50–60 cm in the middle, and 30–40 cm

Fig. 9.54 Surface rupture (red arrows) on the vegetable field at Loc. 139 along the Tanaka-Shimojin segment

Fig. 9.55 Right-lateral offset (40 cm) of the stone wall with south-up separation at Loc. 141 along the Tanaka-Shimojin segment. The white arrows indicate an offset marker

Fig. 9.56 Right-lateral offset (40 cm) of the ditch with south-up separation on the footpath at Loc. 143 along the Tanaka-Shimojin segment. The white and red arrows indicate an offset marker and the surface rupture, respectively

Fig. 9.57 Right-lateral offset (35 cm) of the center joint of the road with south-up separation (30 cm) at Loc. 144. The white and red arrows indicate an offset marker and the surface rupture, respectively

Fig. 9.58 Normal fault ruptures (red arrows) with south-up separation (30 cm) on gravel-filled flat space at Loc. 146

Fig. 9.59 Normal fault rupture (red arrows) with south-up separation (25 cm) on the footpath at Loc. 147

Fig. 9.60 South-up separation (30 cm) on the slope with right-lateral offset (15 cm) based on matched scarp edges at Loc. 150. The white and red arrows indicate an offset marker and the surface rupture, respectively

Fig. 9.61 South-up separation (60 cm) on the slope with right-lateral offset (50 cm) at Loc. 155

Fig. 9.62 Normal fault rupture (red arrows) with south-up separation (60 cm) on slope surface at Loc. 156

Fig. 9.63 Right-lateral offset (30 cm) of road-side concrete wall with south-up displacement (25 cm) at Loc. 157. The white and red arrows indicate an offset marker and the surface rupture, respectively

Fig. 9.64 Normal fault rupture (red arrows) on hill slope with south-up displacement (50 cm) at Loc. 158

Fig. 9.65 Right-lateral offset of the line of potato vegetation (65 cm) with south-up displacement (34 cm) on the vegetable field at Loc. 160. The white and red arrows indicate an offset marker and the surface rupture, respectively

Fig. 9.66 Right-lateral offset of an edge of paddy field with south-up separation at Loc. 160a. The red arrows indicate the surface rupture

Fig. 9.67 Surface rupture (red arrows) with up-on-the-south displacement across the paddy field at Loc. 160a

Fig. 9.68 South-up separation on the hill slope at Loc. 161. The red arrows indicate the surface rupture

in the east. On a temporary outcrop that appeared during the repair work of riverbank southwest of Loc. 160a (Figs. 9.66 and 9.67), Takahashi et al. (2017) found an N74°E-trending and 57°N-dipping fault plane within a fluvial gravel unit dated at 1990–2300 cal BP just beneath the surface rupture of the 2016 earthquake. The gravel unit had a cumulative vertical displacement of ~1.5 m, larger than the coseismic vertical displacement around here (~50 cm). To the northeast, we traced the surface ruptures with south-side-up displacement for ~500 m across the mountain slope (Fig. 9.68).

At Shimojin near the northern termination of the Tanaka-Shimojin segment, an N65°W-trending 600-m-long conjugate rupture appeared. This rupture extended northwest

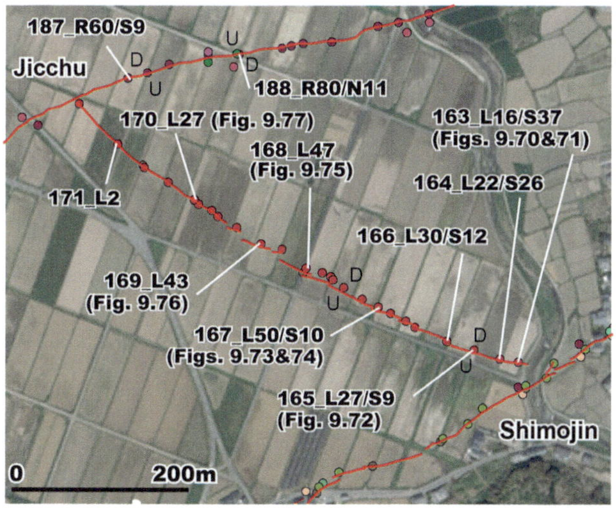

Fig. 9.69 Detailed map of the conjugate surface ruptures between Shimojin and Jicchu. The small circles indicate the measuring points by hand-held GNSS devices, and different colors correspond to different observers. The labels U and D represent the uplifted and the downthrown sides, respectively. The location of the map is shown in Fig. 9.2. The background air photo was taken by the Geospatial Information Authority of Japan

Fig. 9.70 Conjugate rupture (red arrows) across the paddy field at Loc. 163. The location of the photo is shown in Fig. 9.69

Fig. 9.71 Conjugate surface rupture on the paddy field at Loc. 163. The white and red arrows indicate an offset marker and the surface ruptures, respectively. The amount of left-lateral offset of the line of rice stubble was 16 cm. The amount of south-up separation was 37 cm. The location of the photo is shown in Fig. 9.69

to Jicchu, where an N75°E-trending surface rupture appeared on the alluvial plain (Fig. 9.69). This conjugate left-lateral fault was almost straight but consisted of right-stepping *en echelon* fractures (Figs. 9.70, 9.71, 9.72, 9.73, 9.74, 9.75, 9.76, and 9.77). The amount of lateral offset was 40–50 cm in the middle part and decreased to both ends. The

Fig. 9.72 Conjugate surface rupture at Loc. 165. The white and red arrows indicate an offset marker and the surface rupture, respectively. The amount of left-lateral offset of the edge of the concrete road was 27 cm. The amount of south-up separation was 9 cm. The location of the photo is shown in Fig. 9.69

Fig. 9.73 Conjugate surface rupture on the paddy field at Loc. 167. The white and red arrows indicate an offset marker and the surface rupture, respectively. The left-lateral offset of the line of rice stubble was 50 cm, representing the maximum horizontal amount of the conjugate fault. The amount of south-up separation was 10 cm. The location of the photo is shown in Fig. 9.69

Fig. 9.74 Conjugate surface ruptures (red arrows) across the paddy field at Loc. 167. The ruptures showed a right-stepping *en echelon* structure. The location of the photo is shown in Fig. 9.69

up-on-the-south vertical displacement was more prominent to the east and reached a maximum value of 37 cm at Loc. 163.

About 300 m to the northeast from the northern end of the Tanaka-Shimojin segment, we mapped scattered faults 20–120 m long on a hill (Figs. 9.78, 9.79 and 9.80). These faults run parallel each other within a 250-m-wide zone. The fault displacement was up-on-the-north with a maximum of 20 cm, forming uphill-facing scarps.

Fig. 9.75 Conjugate surface rupture at Loc. 168. The white and red arrows indicate an offset marker and the surface ruptures, respectively. The amount of left-lateral offset of the edge of the road was 47 cm. The location of the photo is shown in Fig. 9.69

Fig. 9.76 Conjugate surface rupture on the paddy field at Loc. 169. The white and red arrows indicate an offset marker and the surface rupture, respectively. The amount of left-lateral offset of the tractor rut was 43 cm. The location of the photo is shown in Fig. 9.69

Fig. 9.77 Conjugate surface rupture on the paddy field at Loc. 170. The white and red arrows indicate an offset marker and the surface rupture, respectively. The amount of left-lateral offset of the land boundary was 27 cm. The location of the photo is shown in Fig. 9.69

Fig. 9.78 Surface rupture (red arrows) with north-up separation on the vegetable field at Loc. 171a

Fig. 9.79 Surface rupture (red arrows) of north-up separation (20 cm) on the vegetable field at Loc. 171b

Fig. 9.80 Surface rupture (red arrows) of north-up separation on the slope at Loc. 171c

Acknowledgements This study was supported by the JSPS grants 18H03601.

References

Takahashi N, Ishimura D, Toda S, Nakata T, Watanabe M (2017) Vertical slip rate on a normal fault co-ruptured with the Futagawa fault at the 2016 Kumamoto earthquake: a fault outcrop at Shimojin, Mashiki Town in Kumamoto Prefecture. Active Fault Res 46:27–32

The Research Group for Active Faults of Japan (ed) (1991) Active faults in Japan: sheet maps and inventories (Revised edition). University of Tokyo Press, Tokyo

Yoshioka T, Shintani K, Iemura K, Miyawaki R (2007) Paleoseismicity of the Futagawa-Hinagu fault zone, central Kyushu, Japan. Ann Rep Active Fault Paleoearthquakes Res 7:241–258

Surface Ruptures Along the Central–Northern Part of the Futagawa Fault

10

Yasuhiro Kumahara, Shinji Toda, Hiroyuki Tsutsumi, Hideaki Goto, Daisuke Ishimura, Shinsuke Okada, Kyoko Kagohara, and Heitaro Kaneda

Abstract

This chapter describes the characteristics of the ~24-km-long surface ruptures along the central–northern part of the Futagawa fault during the mainshock of the 2016 Kumamoto earthquake sequence. The rupture appeared from Kiyama in Mashiki Town westward through Nishihara Village, and extended into the Aso caldera up to Kurokawa in Minami-Aso Village. We report the locations and displacements of the ruptures with photographs and detailed maps. The surface ruptures in this section were predominantly right-lateral and generally accompanied vertical displacement up-to-the-south in the southwestern part and up-to-the-north in the northeastern part. At Dozon in Mashiki Town, we measured the largest right-lateral offset of 225 cm among the entire surface ruptures of the 2016 Kumamoto earthquake sequence.

Keywords

Surface rupture • 2016 Kumamoto earthquake • Futagawa fault • Aso caldera

Supplementary Information
The online version contains supplementary material available at (https://doi.org/10.1007/978-981-19-1150-7_10).
Note
The supplementary material of this book contains geographical information system (GIS) and spreadsheet files of the surface ruptures associated with the mainshock of the 2016 Kumamoto earthquake sequence

Y. Kumahara (✉)
Graduate School of Humanities and Social Sciences, Hiroshima University, 1-1-1, Kagamiyama, Higashi-Hiroshima, Hiroshima, 739-8524, Japan
e-mail: kumakuma@hiroshima-u.ac.jp

S. Toda
International Research Institute of Disaster Science (IRIDeS), Tohoku University, 468-1, Aoba, Aramaki, Aoba-ku, Sendai, Miyagi, 980-0845, Japan
e-mail: toda@irides.tohoku.ac.jp

H. Tsutsumi
Department of Environmental Systems Science, Faculty of Science and Engineering, Doshisha University, 1-3, Tatara-Miyakodani, Kyotanabe, Kyoto, 610-0394, Japan
e-mail: htsutsum@mail.doshisha.ac.jp

H. Goto
Graduate School of Humanities and Social Sciences, Hiroshima University, 1-2-3, Higashi-Hiroshima, Hiroshima, 739-8522, Japan
e-mail: hgoto@hiroshima-u.ac.jp

D. Ishimura
Department of Geography, Tokyo Metropolitan University, 1-1, Minami-Osawa, Hachioji, Tokyo, 192-0397, Japan
e-mail: ishimura@tmu.ac.jp

S. Okada
Faculty of Science and Engineering, Iwate University, 4-3-5, Ueda, Morioka, Iwate, 020-8551, Japan
e-mail: sokada@iwate-u.ac.jp

K. Kagohara
Faculty of Education, Yamaguchi University, 1677-1, Yoshida, Yamaguchi, Yamaguchi, 753-8513, Japan
e-mail: k-kago@yamaguchi-u.ac.jp

H. Kaneda
Department of Civil and Environmental Engineering, Chuo University, 1-13-27, Kasuga, Bunkyo-ku, Tokyo, 112-8551, Japan
e-mail: kaneda.00c@g.chuo-u.ac.jp

The 2016 surface ruptures from Kiyama in Mashiki Town to Kurokawa in Minami-Aso Village along the central–northern Futagawa fault were ~24 km long and almost continuous. We divided the ruptures into five subsections based on

Y. Kumahara et al. (eds.), *Surface Ruptures Associated with the 2016 Kumamoto Earthquake Sequence in Southwest Japan*, Advances in Geological Science, https://doi.org/10.1007/978-981-19-1150-7_10

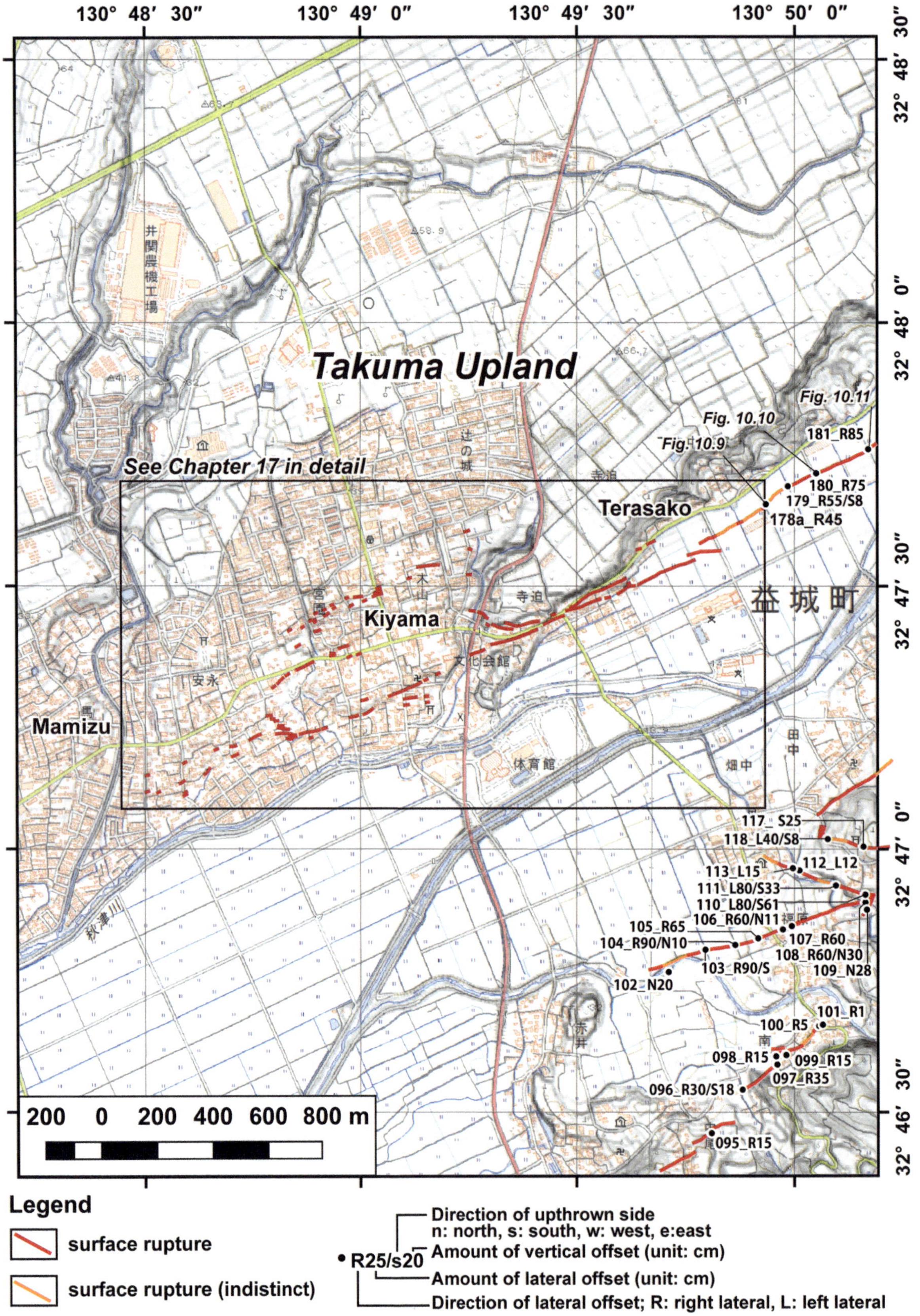

Fig. 10.1 Detailed surface rupture map of the central–northern part of the Futagawa fault around Kiyama, Mashiki Town. Strike-slip and vertical displacements at each locality are also shown. The background is a half-transparent topographic map of Geospatial Information Authority of Japan (GSI) that is underlain by a slope map produced from 5 m grid digital elevation models (DEMs) of GSI. The surface ruptures around Kiyama are described in detail in Chap. 17

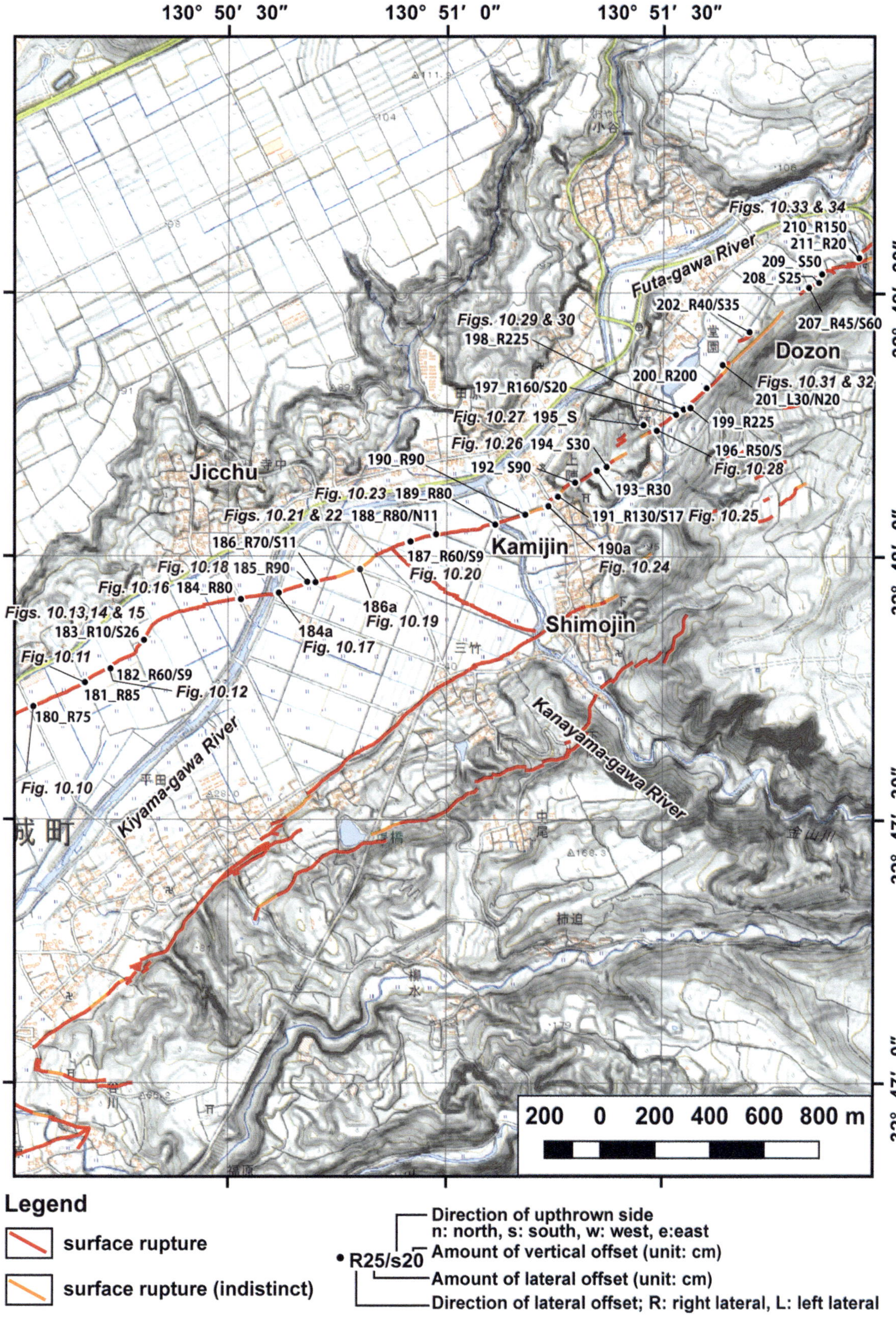

Fig. 10.2 Detailed surface rupture map of the central–northern part of the Futagawa fault around Jicchu and Dozon, Mashiki Town. Strike-slip and vertical displacements at each locality are also shown. The base map is the same as in Fig. 10.1

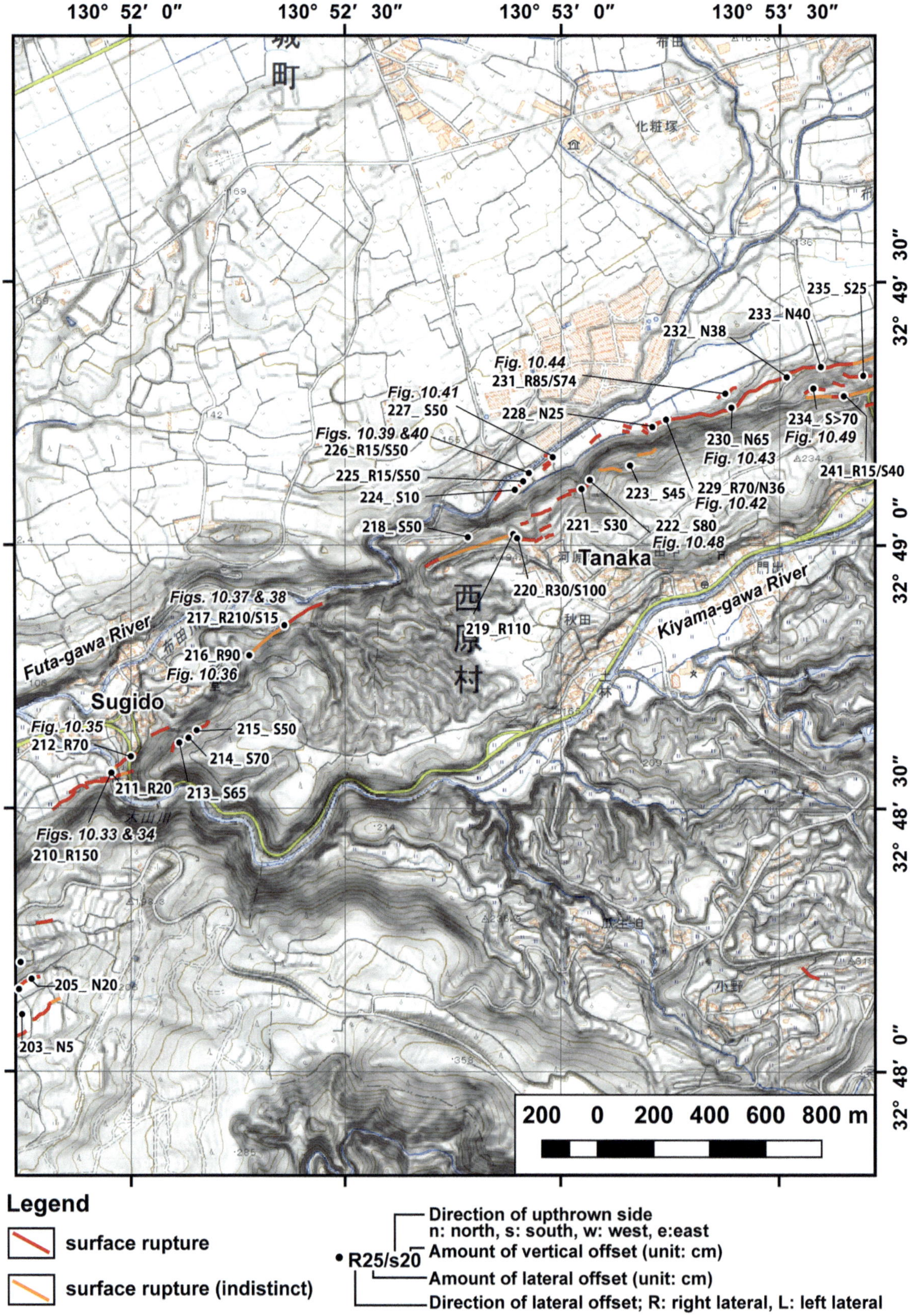

Fig. 10.3 Detailed surface rupture map of the central–northern part of the Futagawa fault around Sugido, Mashiki Town, and Tanaka, Nishihara Village. Strike-slip and vertical displacements at each locality are also shown. The base map is the same as in Fig. 10.1

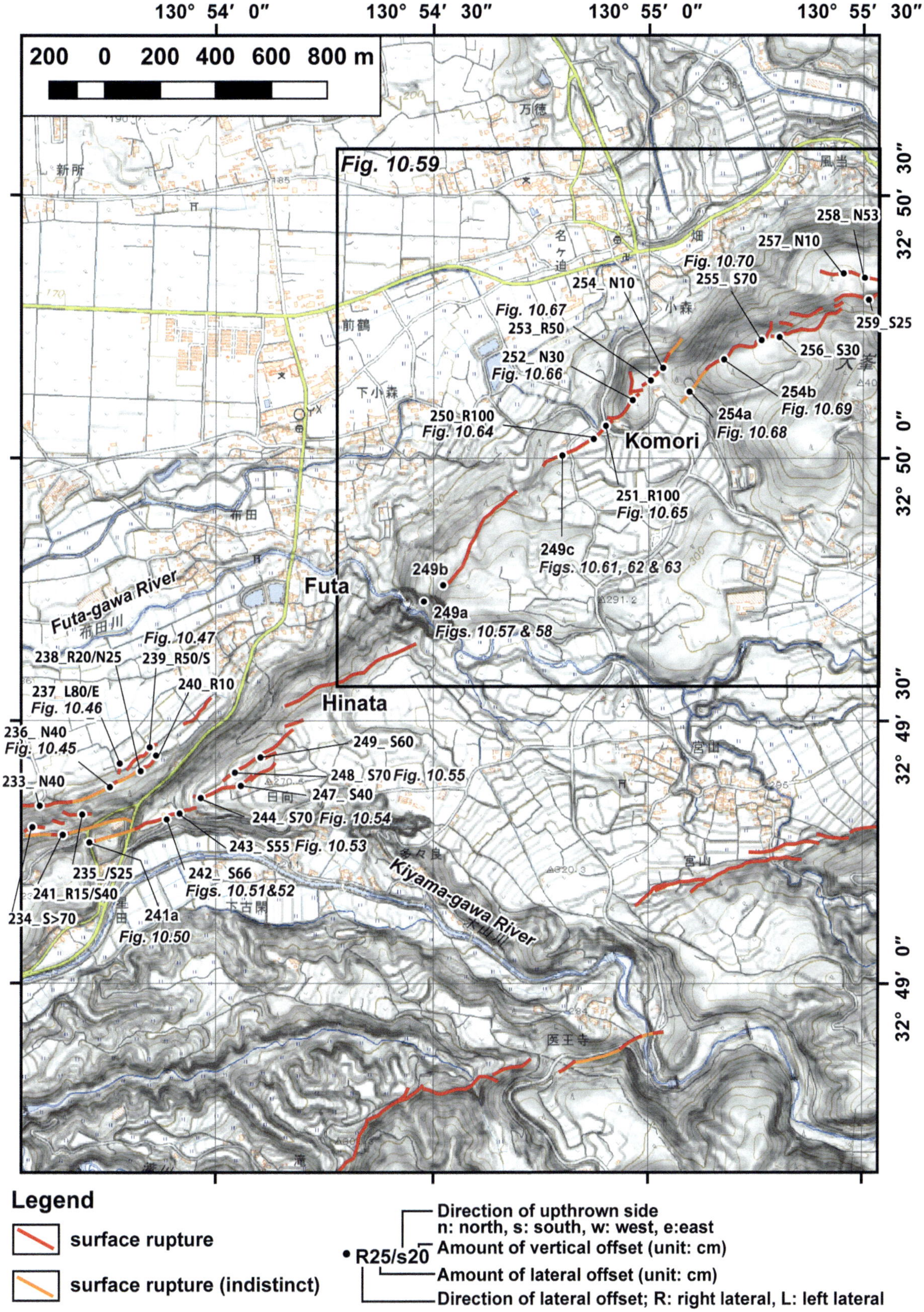

Fig. 10.4 Detailed surface rupture map of the central–northern part of the Futagawa fault around Futa, Nishihara Village. Strike-slip and vertical displacements at each locality are also shown. The base map is the same as in Fig. 10.1

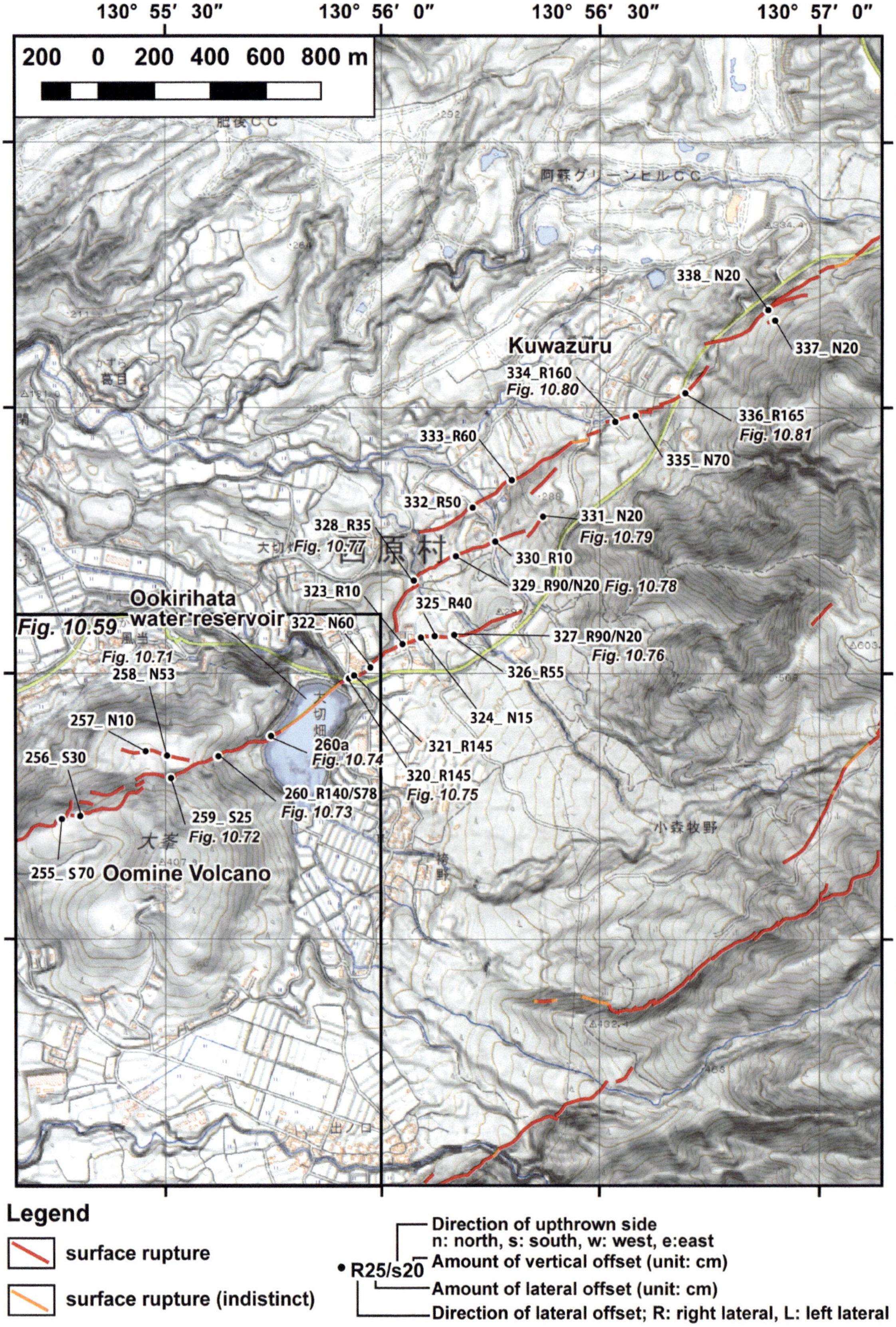

Fig. 10.5 Detailed surface rupture map of the central–northern part of the Futagawa fault around Oomine and Kuwazuru, Nishihara Village. Strike-slip and vertical displacements at each locality are also shown. The base map is the same as in Fig. 10.1

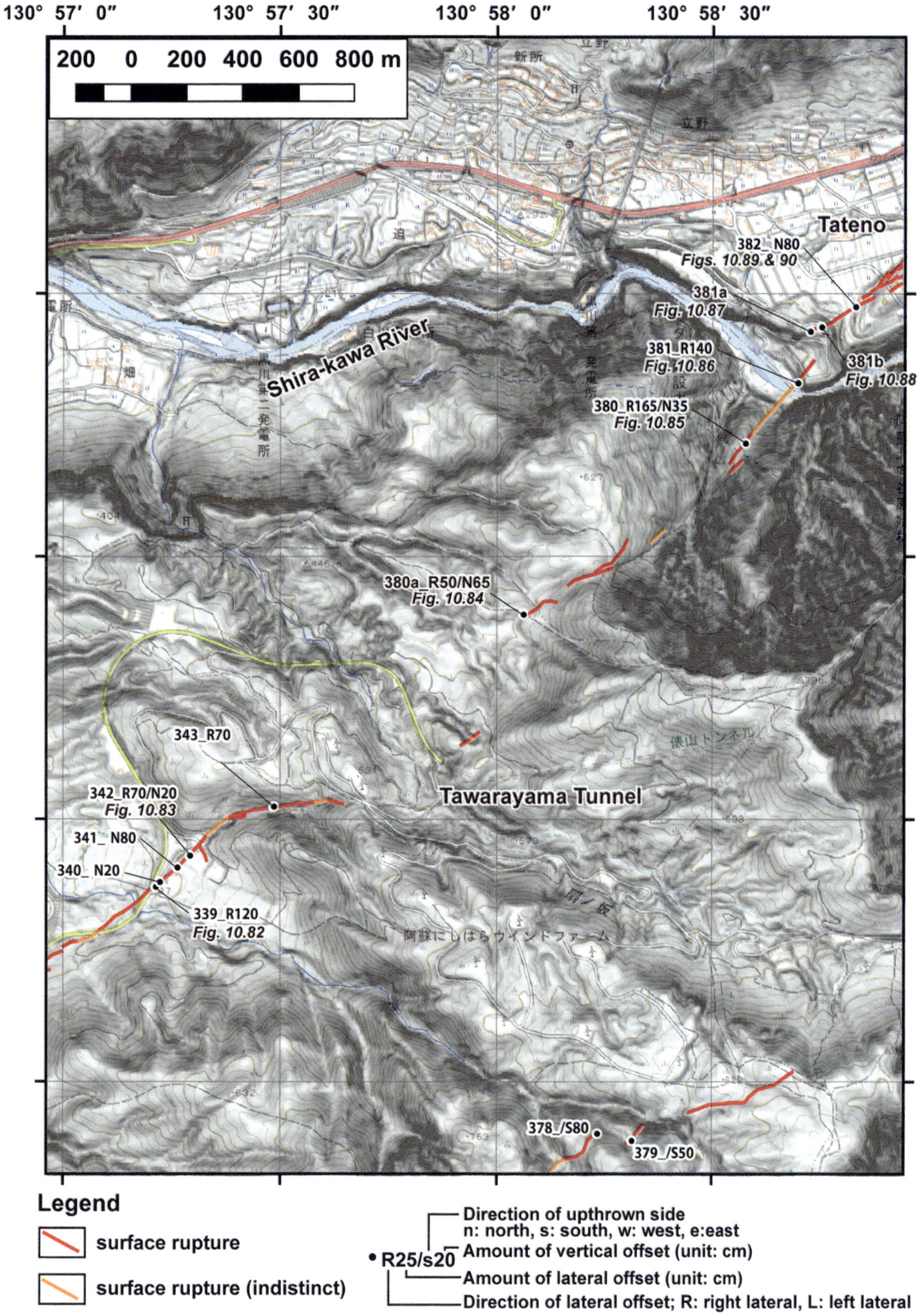

Fig. 10.6 Detailed surface rupture map of the central–northern part of the Futagawa fault around Tawarayama, Nishihara Village, and Tateno, Minami-Aso Village. Strike-slip and vertical displacements at each locality are also shown. The base map is the same as in Fig. 10.1

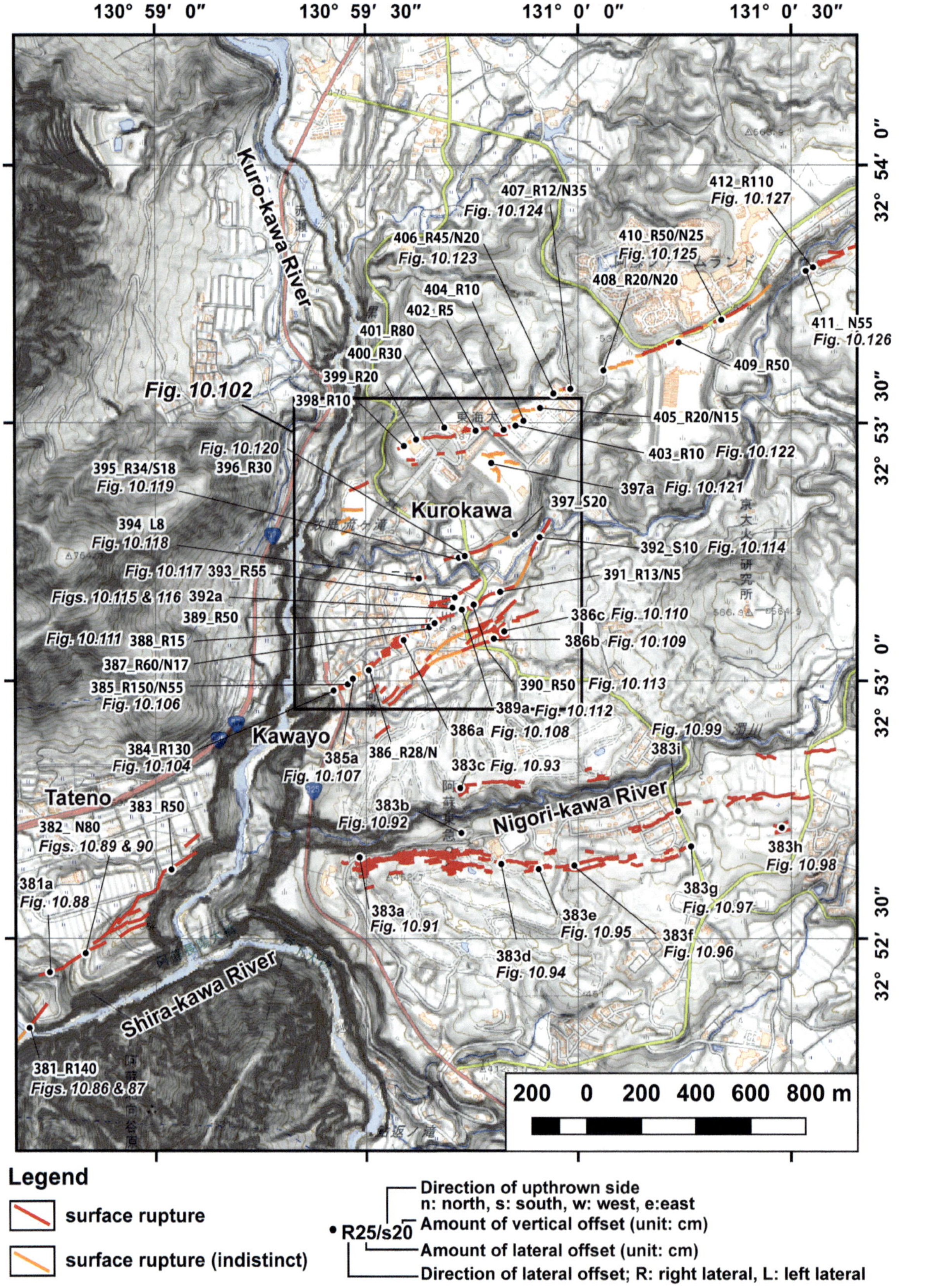

Fig. 10.7 Detailed surface rupture map of the central–northern part of the Futagawa fault around Kawayo and Kurokawa, Minami-Aso Village. Strike-slip and vertical displacements at each locality are also shown. The base map is the same as in Fig. 10.1

Fig. 10.8 Detailed surface rupture map of the central–northern part of the Futagawa fault east of Kurokawa, Minami-Aso Village. Strike-slip and vertical displacements at each locality are also shown. The base map is the same as in Fig. 10.1

their geomorphic characteristics: (1) along the base of the south-facing scarp that bounds the southern margin of the Takuma Upland (Fig. 10.1), (2) on the alluvial plain of the Kiyama River (Fig. 10.2), (3) along the northwestern margin of the hills composed of the pyroclastic flow deposits or lavas (Figs. 10.3, 10.4, and 10.5), (4) across the outer rim of the Aso caldera (Fig. 10.6), and (5) inside the Aso caldera (Figs. 10.7 and 10.8). The rupture in the middle part closely followed the active Futagawa fault, which had been mapped in detail by Ikeda et al. (2001). The right-lateral slip was predominant along this section, with the offset ranging from a few to 225 cm. The vertical displacement along the rupture was generally up-on-the-south in the southwestern part and up-on-the-north in the northeastern part. The surface ruptures in the southernmost part of this section in downtown Mashiki will be described in detail in Chap. 17, in which they report widely distributed co-seismic cracks that severely damaged infrastructures and houses and examine their relations to tectonic landforms.

The southwestern end of the fault in this section laid at Kiyama in downtown Mashiki at the base of a south-facing scarp of the Takuma Upland composed of the ~90-ka Aso-4 pyroclastic flow deposits (Fig. 10.1). From there, a continuous trace extended northeastward. From Terasako to Kamijin (Figs. 10.1 and 10.2), the surface rupture appeared on the alluvial plain of the Kiyama River, north of and subparallel to the previously mapped Futagawa fault (Figs. 10.9, 10.10, 10.11, 10.12, 10.13, 10.14, 10.15, 10.16, 10.17, 10.18, 10.19, 10.20, 10.21, 10.22, 10.23, 10.24, and 10.25). The rupture appeared as a single straight trace, and the amount of right-lateral slip increased toward the northeast from 45 cm at Loc. 178a (Fig. 10.9) to 130 cm at Loc. 191 (Fig. 10.25). The vertical displacement was not prominent except around Loc. 183. The rupture around Loc. 183 trended N30°E, locally rotated counterclockwise ~30° from the general trend, forming a restraining bend for the right-lateral strike-slip fault. Here, we observed a thrust fault structure with a 26 cm south-side-up vertical displacement, whereas the right-lateral slip was much smaller (10 cm) than the neighboring sections (Fig. 10.13). Tsutsumi et al. (2018) conducted a trenching survey at Loc. 183 (Figs. 10.14 and 10.15) and exposed south-dipping reverse faults that deformed fluvial deposits derived from the Kiyama River and Aso-4 pyroclastic flow deposits (~90 ka). Vertical separation of the fluvial deposits and Aso-4 pyroclastic flow deposits was more than 2 m, suggesting that reverse faulting

Fig. 10.9 Photograph of right-lateral offset (amount 45 cm) of a straight white line on the road at Loc. 178a. White and red arrows indicate the offset marker and surface rupture, respectively

Fig. 10.10 Photograph of right-lateral offset (amount 75 cm) of straight irrigation beside the road at Loc. 180. White and red arrows indicate the offset marker and surface rupture, respectively

Fig. 10.11 Photograph of right-lateral offset (amount 85 cm) of straight irrigation beside the road at Loc. 181. White and red arrows indicate the offset marker and surface rupture, respectively

Fig. 10.12 Photograph of right-lateral offset (amount 60 cm) of a straight concrete boundary of the paddy field with minor south-up separation (9 cm) at Loc. 182. White and red arrows indicate the offset marker and surface rupture, respectively

Fig. 10.13 Photograph of compressional south-up vertical separation (amount 26 cm) on the paddy field with right-lateral offset (amount 10 cm) at Loc. 183. The rupture shows a kind of "mole track" structure. The black dashed line box indicates the location of the trench site (Tsutsumi et al. 2018)

occurred on the same trace repeatedly in the late Quaternary. Between Kamijin and Dozon, the ruptures traversed a hill composed of Mid-Quaternary pyroclastic flow deposits from the Aso Volcano. On top of the hill, the rupture was scattered (Fig. 10.26). East of the hilltop, the rupture was split into two parallel traces. The northern trace showed a south-side-up displacement at Loc. 195 (Fig. 10.27), and the southern trace showed a 50 cm right-lateral slip at Loc. 196 (Fig. 10.28).

The surface ruptures in Dozon appeared along the northern foot of the hill. The right-lateral offsets of the footpaths at Locs. 198 and 199 were as large as 225 cm, the maximum strike-slip displacement on the 2016 surface ruptures (Figs. 10.29 and 10.30). Around Dozon, we also identified small-scale surface deformation such as flexure on cultivated land as well as an N80°E-trending conjugate fault. The offset along the conjugate rupture was left-lateral (30 cm) with south-side-up vertical separation (20 cm) at

Fig. 10.14 Photograph of the west wall of the trench across the surface rupture at Loc. 183 (Tsutsumi et al. 2018). Red arrows indicate the fault on the trench wall

Fig. 10.15 Photograph of the east wall of the trench at Loc. 183 (Tsutsumi et al., 2018). The size of the yellow-string grid is 1 m by 1 m. Red arrows indicate the fault plane. White and blue arrows show the vertical separation (26 cm) of the 2016 earthquake and the larger, cumulative vertical separation of the top of a gravel layer, respectively. The fault dips southeast at 45–60°

Fig. 10.16 Photograph of right-lateral offset (amount 80 cm) of a straight road edge at Loc. 184. White and red arrows indicate the offset marker and surface ruptures, respectively

Fig. 10.17 Photograph of right-lateral offset at Loc. 184a. Red arrows indicate the surface rupture

Fig. 10.18 Photograph of right-lateral offset (amount 90 cm) of a rice paddy dike at Loc. 185. White and red arrows indicate the offset marker and surface rupture, respectively

Fig. 10.19 Photograph of right-lateral offset of the center line on the road at Loc. 186a. White and red arrows indicate the offset marker and surface rupture, respectively

Fig. 10.20 Photograph of right-lateral offset (amount 60 cm) of the lines of rice stubble with up-on-the-south vertical displacement (amount 9 cm) at Loc. 187. White and red arrows indicate the offset marker and surface rupture, respectively

Fig. 10.21 Photograph of surface rupture on paddy field with left-stepping *en echelon* structure at Loc. 188. Red arrows indicate the surface rupture

Fig. 10.22 Photograph of right-lateral offset (amount 80 cm) of a straight edge of paddy field with minor north-up separation (amount 11 cm) at Loc. 188. White and red arrows indicate the offset marker and surface rupture, respectively

Fig. 10.23 Photograph of right-lateral offset (amount 80 cm) of a straight road edge at Loc. 189. White and red arrows indicate the offset marker and surface rupture, respectively

Fig. 10.24 Photograph of right-lateral offset of a gutter at Loc. 190a. White and red arrows indicate the offset marker and surface rupture, respectively. The photo was taken on the day of the mainshock

Fig. 10.25 Photograph of right-lateral offset (amount 130 cm) of straight stone lines with minor south-up vertical separation (amount 17 cm) at Loc. 191. White and red arrows indicate the offset marker and surface rupture, respectively

Fig. 10.26 Photograph of surface ruptures (red arrows) on the hill composed of pyroclastic flow deposits at Loc. 194. Note that all the tombstones toppled down

Fig. 10.27 Photograph of south-up vertical deformation (amount over 50 cm) on the road at Loc. 195. Red arrows indicate the surface rupture

Fig. 10.28 Photograph of right-lateral offset (amount 50 cm) of a white line on the road at Loc. 196. White and red arrows indicate the offset marker and surface rupture, respectively

Loc. 201 (Fig. 10.31). Iwasa et al. (2020) conducted a trench survey across the conjugate rupture. They identified more than 40 cm vertical displacement of strata on the trench walls (Fig. 10.32), indicating that even the conjugate fault moved repeatedly. For a distance of 400 m east of Loc. 200, the precise location of the main surface rupture was unclear because it passed through the forest. At Locs. 210 and 211, two traces ran parallel with a left step (Fig. 10.33). It was not easy to follow the trace continuously in the dense forest between Loc. 211 and Loc. 219, but we identified right-lateral offsets at Locs. 210 (Fig. 10.34), 216 and 217. The amounts of right-lateral offsets were 150 cm at Loc. 210 (Fig. 10.33), 70 cm at Loc. 212 (Fig. 10.35), 90 cm at Loc. 216 (Fig. 10.36), and 210 cm at Loc. 217 (Figs. 10.37

Fig. 10.29 Photograph of right-lateral offset of a paddy dike at Loc. 198 with the largest amount of offset (225 cm) associated with the 2016 Kumamoto earthquake. White and red arrows indicate the offset marker and surface rupture, respectively

Fig. 10.30 Photograph of surface rupture on the farmland at Loc. 198. Red arrows indicate the fault trace

Fig. 10.31 Photograph of left-lateral offset (amount 30 cm) of a gutter on an N80°E-trending conjugated rupture with north-up vertical separation (amount 20 cm) at Loc. 201. White and red arrows indicate the offset marker and surface rupture, respectively

Fig. 10.32 Photograph of the west wall of the trench across the conjugated fault at Loc. 201 (Iwasa et al. 2020). The interval between vertical yellow strings is 1 m. Red arrows indicate the fault planes. Because the survey was conducted one year after the earthquake, the surface rupture was already vanished by tractor plowing. Strata on the wall showed larger vertical displacements (>40 cm) than that due to the 2016 earthquake (20 cm)

Fig. 10.33 Photograph of parallel surface ruptures with a left-stepping structure at Locs. 210 and 211. Red arrows indicate the surface ruptures

Fig. 10.34 Photograph of right-lateral offset (amount 150 cm) of an edge of the concrete road at Loc. 210. White and red arrows indicate the offset marker and surface rupture, respectively

Fig. 10.35 Photograph of right-lateral offset (amount 70 cm) of a white line on the road at Loc. 212. White and red arrows indicate the offset marker and surface rupture, respectively

and 10.38). On the hillside slope south of the main trace, a short trace with south-side-up vertical displacement was observed at Locs. 213, 214 and 215. In the ~500-m-long section between east of Loc. 217 at Sugido and west of Loc. 218 at Tanaka, the location of the main surface rupture was unknown due to a large landslide.

The rupture from Tanaka to Hinata was 3.3 km long and principally consisted of two parallel traces: the northern trace along or close to the base of the 90-m-high fault scarp that bounds the northern margin of the hill and the southern trace near the scarp crest (Figs. 10.3 and 10.4). From Locs. 224 to 228 along the northern trace, several ~50-m-long ruptures appeared ~20 m away from the base of the fault scarp. We identified ~50 cm south-up vertical displacement with ~15 cm right-lateral offset at Locs. 225 (Figs. 10.39 and 10.40), 226 and 227 (Fig. 10.41). From Loc. 228 to the west of Hinata along the northern trace, the rupture emerged continuously along the scarp base (Figs. 10.3 and 10.4). The

Fig. 10.36 Photograph of right-lateral offset (amount 90 cm) of an edge of the concrete road at Loc. 216. White and red arrows indicate the offset marker and surface rupture, respectively

Fig. 10.37 Photograph of right-lateral offset (amount 210 cm) of the steps to the Shioi Shrine at Loc. 217. A large holy tree on the rupture fell down. White and red arrows indicate the offset marker and surface rupture, respectively

Fig. 10.38 Photograph of right-lateral offset (amount 210 cm) of the step to the Shioi Shrine at Loc. 217. White and red arrows indicate the offset marker and surface rupture, respectively

measured vertical displacements were ~40 cm up-to-the-north, implying that the co-seismic breaks did not contribute to the growth of the fault scarp (Figs. 10.42, 10.43 and 10.45). However, we locally observed south-up vertical displacement at Locs. 231 (Fig. 10.44) and 239 (Fig. 10.47). We also measured right-lateral slips at several places, and the slip amount was 70 cm at Loc. 229 (Fig. 10.42), 85 cm at Loc. 231 (Fig. 10.44), and 50 cm at Loc. 239 (Fig. 10.47). An N23°E-trending trace, rotated 40 degrees counterclockwise from the general strike of the Futagawa fault, was observed for a length of 40 m at Loc. 237 (Fig. 10.46). Here we measured 80 cm left-lateral offset of tractor ruts with east-up vertical separation. The northern trace terminated west of Hinata.

It was not easy to follow the southern trace continuously because it passed through the dense forest on the fault

Fig. 10.39 Photograph of south-up vertical deformation (amount 50 cm) with right-lateral offset (amount 15 cm) of straight ridges in a sweet potato field at Loc. 225. Red arrows indicate the surface rupture

Fig. 10.40 Photograph of south-up vertical displacement (amount 50 cm) with right-lateral offset (amount 15 cm) of straight ridges in a sweet potato field at Loc. 225. White and red arrows indicate the offset marker and surface rupture, respectively

Fig. 10.41 Photograph of south-up vertical deformation (amount 50 cm) on the vegetable field at Loc. 227. Red arrows indicate the surface rupture

Fig. 10.42 Photograph of right-lateral slip of grass lines (amount 70 cm) with north-up vertical displacement (amount 36 cm) at Loc. 229. White and red arrows indicate the offset marker and surface rupture, respectively

Fig. 10.43 Photograph of north-up vertical displacement (amount 65 cm) on forested slope at Loc. 230

Fig. 10.44 Photograph of right-lateral slip of the boundary of the vegetable fields (amount 85 cm) with south-up vertical displacement (amount 74 cm) at Loc. 231. White and red arrows indicate the offset marker and surface rupture, respectively

Fig. 10.45 Photograph of north-up vertical displacement (amount 40 cm) on the hill slope at Loc. 236. Red arrows indicate the surface rupture

Fig. 10.46 Photograph of conjugate left-lateral offset (amount 80 cm) of track ruts on the vegetable field at Loc. 237. White and red arrows indicate the offset marker and surface rupture, respectively

Fig. 10.47 Photograph of right-lateral slip of an edge of the road (amount 50 cm) with south-up vertical displacement at Loc. 239. White and red arrows indicate the offset marker and surface rupture, respectively

Fig. 10.48 Photograph of south-up vertical displacement (amount 80 cm) on the slope at Loc. 222

Fig. 10.49 Photograph of south-up vertical deformation (amount > 70 cm) on the orchard field at Loc. 234. Red arrows indicate the surface rupture

scarp. We identified open cracks with south-up vertical separation from 30 to 80 cm (Figs. 10.48, 10.49, 10.51, 10.52, 10.53, 10.54, and 10.55). At Loc. 241a, a minor conjugate left-lateral fault with north-up vertical separation was observed (Fig. 10.50). The southern trace consisted of several left-stepping segments and extended northeast.

Along the right bank immediately downstream of the Futa Fall (Loc. 249a), a 220-m-wide and 70-m-high outcrop appeared due to a landslide triggered by the mainshock (Figs. 10.56, 10.57 and 10.58; Ishimura, 2019). This outcrop exposed, from the bottom, the Takayubaru lava, gravel deposits 1, Aso-4 pyroclastic flow deposits, gravel deposits 2, yellowish-brown soil, and black soil (Fig. 10.58; Ishimura, 2019). The Takayubaru lava flowed from the Omine Volcano and formed the Takayubaru Highland (Fig. 10.56). The lava erupted between the Aso-3 (120–135 ka; Machida

Fig. 10.50 Photograph of north-up vertical deformation with a minor left-lateral slip of the white line on the road at Loc. 241a. White and red arrows indicate the offset marker and surface rupture, respectively

Fig. 10.51 Photograph of south-up vertical displacement (amount 66 cm) on the vegetable field at Loc. 242. Red arrows indicate the surface ruptures

Fig. 10.52 Photograph of south-up vertical displacement (amount 66 cm) on the vegetable field at Loc. 242. Red arrows indicate the surface rupture

and Arai 2003) and Aso-4 ($\sim$87 ka; Aoki, 2008) pyroclastic flows. The lava's K–Ar age is 90 $\pm$ 4 ka (Matsumoto et al. 1991) and 81 $\pm$ 4 ka and 98 $\pm$ 18 ka (Miyoshi et al. 2013). At this location, Watanabe and Ono (1969) inferred the presence of the Futagawa fault based on the vertical displacement of the Takayubaru lava and the deformation of the Takayubaru Highland and suggested that the fault was active in the late Pleistocene. On the outcrop, Ishimura (2019) identified several active faults (f1–f6 in Fig. 10.58) displacing the lavas and strata of Quaternary age. Significant

Fig. 10.53 Photograph of south-up vertical displacement (amount 55 cm) in the garden at Loc. 243. Red arrows indicate the surface rupture

Fig. 10.54 Photograph of south-up vertical displacement (amount 70 cm) on the concrete path at Loc. 244

Fig. 10.55 Photograph of south-up vertical displacement (amount 70 cm) on the hill slope at Loc. 248. Red arrows indicate the surface rupture

vertical separation was observed across f2 and f3, indicating that these are the principal faults of the Futagawa fault. Other faults with ~10 m vertical separation are subsidiary faults of the Futagawa fault. We identified fresh open cracks on the ground surface above f1 and f2-1 and confirmed that these faults moved during the 2016 earthquake. Subsidiary normal faults (f5 and f6) formed a small horst (Fig. 10.58). At Loc. 249b, left-stepping cracks were observed along the active fault scarps. Additionally, we observed north-up surface ruptures on the extension of the subsidiary faults (f5) exposed at Loc. 249a, indicating that these subsidiary faults also moved repeatedly and contributed to the development of tectonic geomorphology. Ishimura (2019) estimated both vertical and horizontal slip rates of the Futagawa fault at

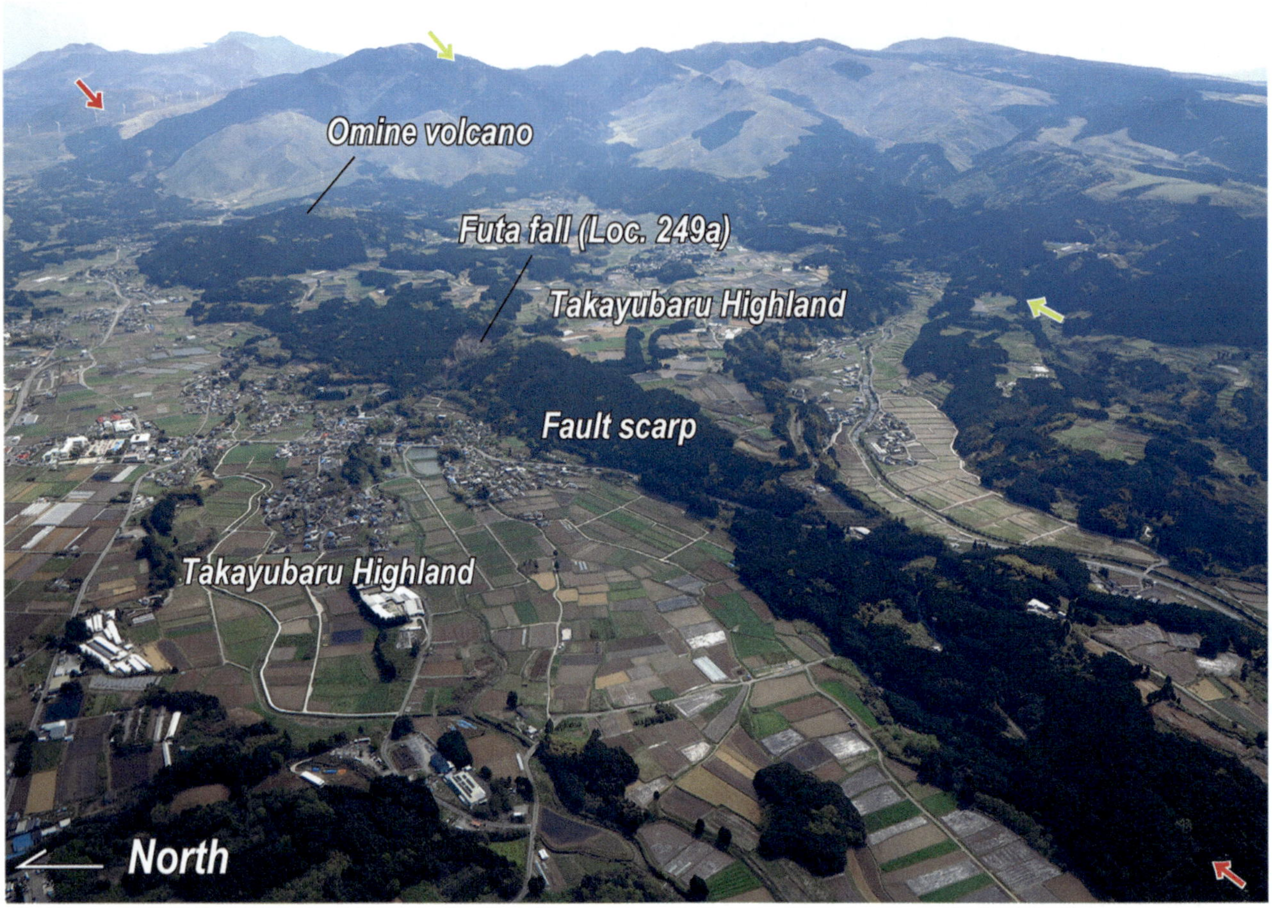

Fig. 10.56 Oblique air photo of the Futagwa fault scarp around Nishihara Village. Red and yellow arrows indicate the Futagawa and Idenokuchi faults, respectively. The Takayubaru lava erupted from the Omine Volcano formed the Takayubaru Highland in the foreground. The photo was taken by Akihiro Murata, Professor Emeritus of Tokushima University

1 mm/year or larger based on displacements and ages of geomorphic surfaces and the Takayubaru lava. The recurrence interval was estimated to be one to two thousand years based on the slip rate and co-seismic displacements of the 2016 event. The estimated slip rate and recurrence interval of the Futagawa fault are higher than the previous estimation and suggest that the Futagawa fault is one of the most active faults in Kyushu region.

The Futagawa fault offsets stream channels right-laterally by 70–150 m around the Omine Volcano (Fig. 10.59). The surface ruptures passed through the locations where the valleys are deflected. In Komori, two trenches were dug across the surface rupture at Loc. 249c after the earthquake (big and small trenches in Fig. 10.60; Iwasa et al. 2022). A ~40-cm-deep elongated depression appeared during the earthquake at this locality (Fig. 10.61). On the trench walls, volcanic soil layers were deformed more than the ground surface (Figs. 10.62 and 10.63), suggesting that a similar depression had appeared during the past faulting events. We interpreted the depression as a small-scale pull-apart basin along the strike-slip fault because the right-lateral offset was dominant along the surface ruptures around the trench site, and there were two flexure-like structures related to the depression on the small trench walls. On the ~700-m-long trace from Loc. 249c to the northeast, we observed right-lateral offset ranging from 50 to 100 cm at Locs. 250 (Fig. 10.64), 251 (Fig. 10.65), and 253 (Fig. 10.67). We also identified up-on-the-north vertical separation of 10–30 cm at Locs. 252 (Fig. 10.66) and 254.

The rupture trace stepped 150 m to the right at Komori and extended northeast, cutting through the northern slope of the cone-shaped Omine Volcano (Fig. 10.59). There were few offset references inside the natural forest, but we were able to identify right-lateral and up-on-the-south vertical displacements at several places (Figs. 10.68, 10.69, 10.70, 10.71, 10.72 and 10.73). At Loc. 260, we measured ~140 cm right-lateral and ~78 cm of up-on-the-south vertical displacement of the footpath, the largest amount in the vicinity (Fig. 10.73). On the 300-m-long and N80° W-trending minor trace ~50 m north of the main trace,

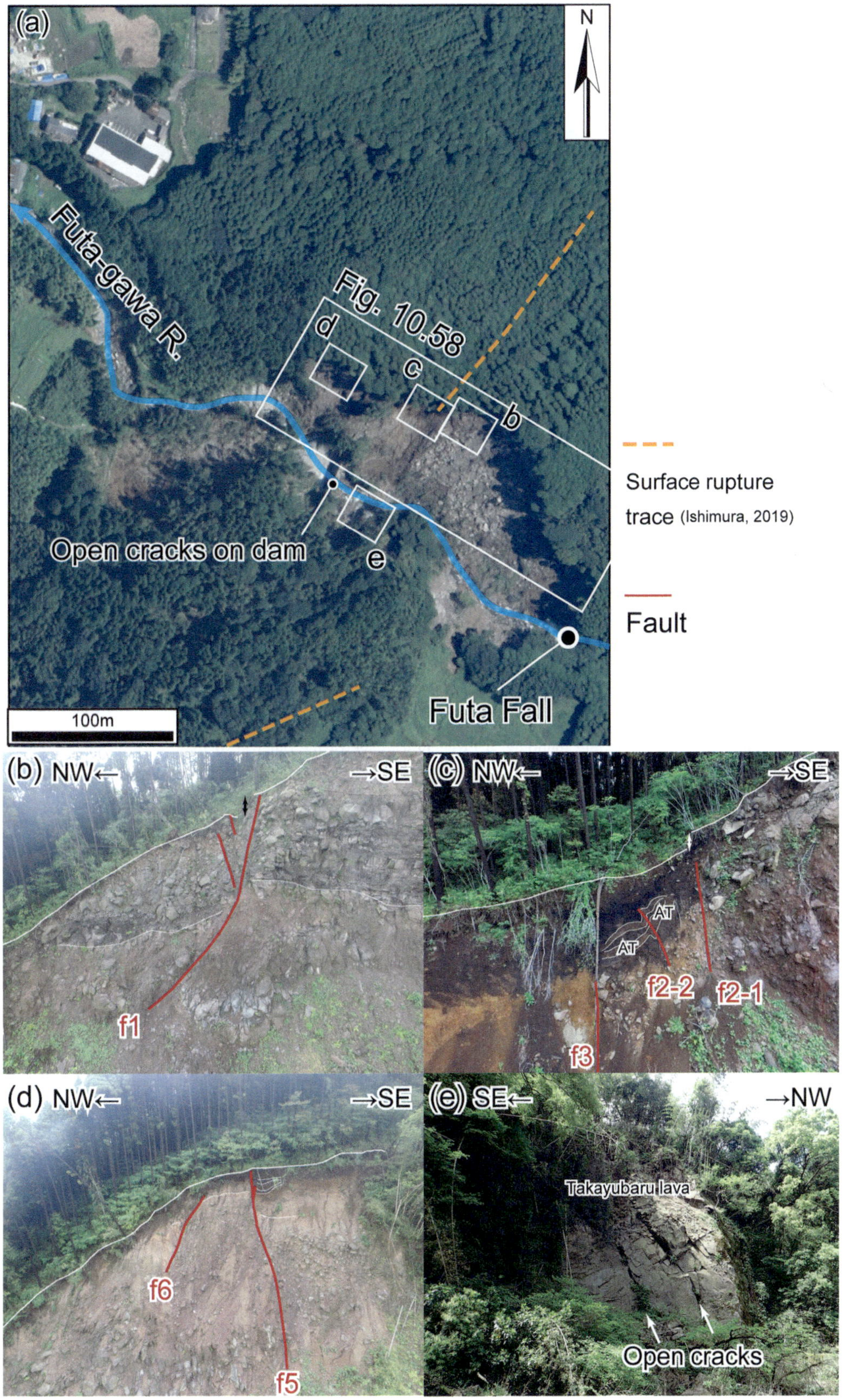

Fig. 10.57 Location and photographs of the outcrop along the Futa River at Locs. 249a and 249b. **a** Aerial photograph after the mainshock provided by Geospatial Information Authority of Japan (GSI). **b–e** Close-up photographs of the outcrop. AT: Aira-Tn tephra (~30 ka; Smith et al. 2013). Modified from Ishimura (2019)

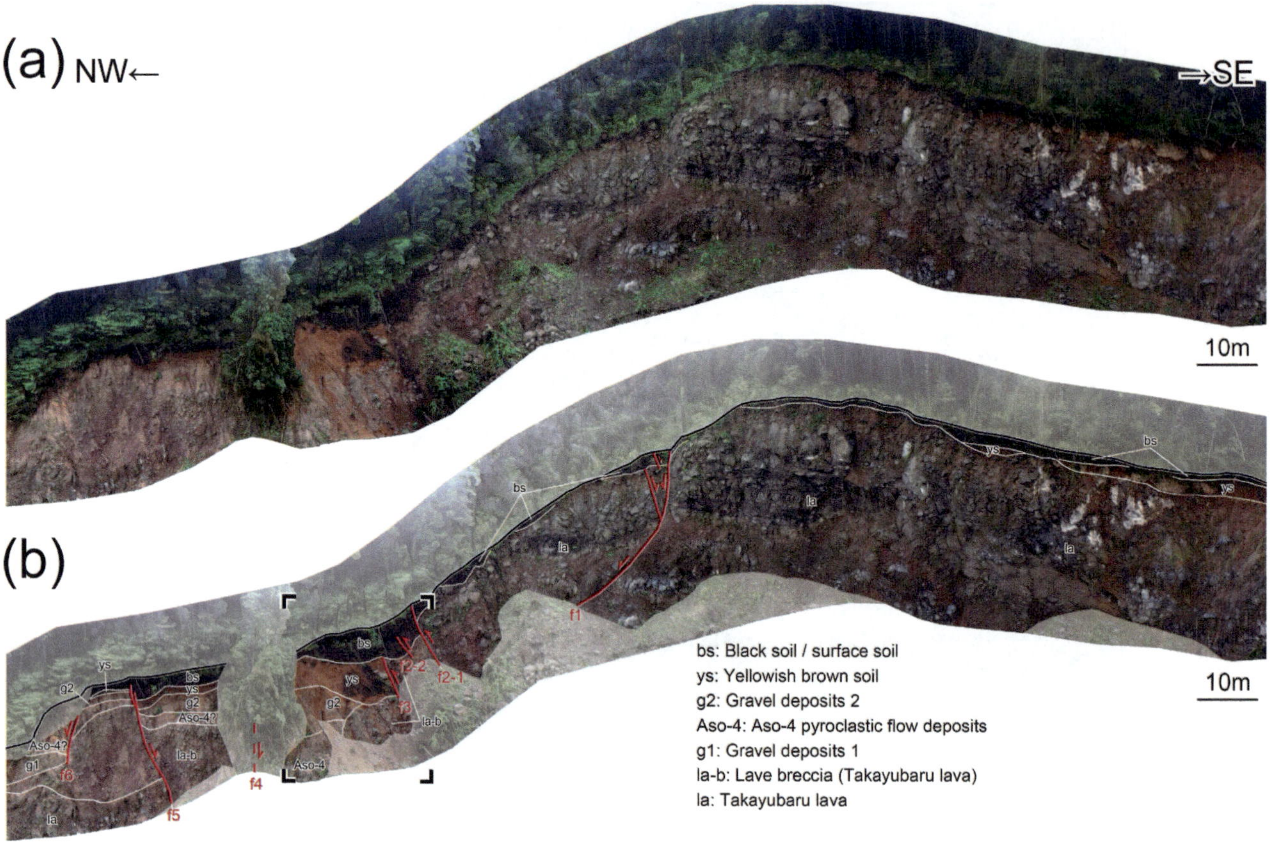

Fig. 10.58 Image of the outcrop based on texture mapping. **a** Projection image of the outcrop. **b** Interpretation of the outcrop. The rectangle shows the area where Ishimura (2019) conducted a detailed field survey. From Ishimura (2019)

we observed up-on-the-north vertical displacement at Locs. 257 and 258 (Fig. 10.71). A depression developed between the main and minor traces. Assuming the volcano's shape was purely conical, we could measure ~150 m right-lateral and ~80 m down-to-the-north vertical displacement along the main trace (Fig. 10.59). This indicates that the volcano has been faulted repeatedly during past earthquakes similar to the 2016 event.

The 3.3-km-long surface ruptures from the Ookirihata water reservoir to the Tawarayama Tunnel crossed the northward-sloping fan surfaces and mountain slopes (Figs. 10.5 and 10.6). The overall geometry of the rupture traces exhibited a left-stepping *en echelon* pattern. Right-lateral offsets were recognized at several points (Figs. 10.74, 10.75, 10.76, 10.77 and 10.78, 10.80, 10.81, and 10.82) with the maximum displacement being ~165 cm at Loc. 336 (Fig. 10.81). The vertical displacement on this section was up-on-the-north (20–70 cm), forming the uphill-facing scarps (Figs. 10.78, 10.79 and 10.83). From the Tawarayama Tunnel to the northeast, it was not easy to follow the trace because of the thick vegetation. Nonetheless, we identified a minor left-stepping surface rupture in the forest immediately northeast of the tunnel entrance, and this rupture trace appeared to die out around here.

Before the earthquake, no study suggested that the Futagawa fault extends into the Aso caldera, although the rivers that originate in the caldera are all drained from where the fault hits the caldera rim (Fig. 1.2 in Chap. 1). In 2016, the surface rupture crossed the outer rim and extended well into the caldera. After making a ~300-m-wide left step, the surface rupture reappeared at Loc. 380a, where the fault cut across an old dirt road and formed a ~65-cm-high south-facing monoclinal scarp that is also associated with ~50 cm right-lateral displacement and large open cracks (Fig. 10.84). The rupture then crossed a small pass and went down toward the Shirakawa River, an only drainage from the Aso caldera. At Loc. 380, the surface rupture extended following a small valley with north-up vertical displacement (Fig. 10.85). On the right bank of the Shirakawa River, there was a fault outcrop cutting the Tateno lava at Loc. 381 (Fig. 10.86). The fault plane was almost vertical, and the right-lateral offset was ~140 cm. The surface rupture offsets the train rail right-laterally at Loc. 381a (Fig. 10.87). At Loc. 381b, several right-lateral ruptures showed a left-stepping

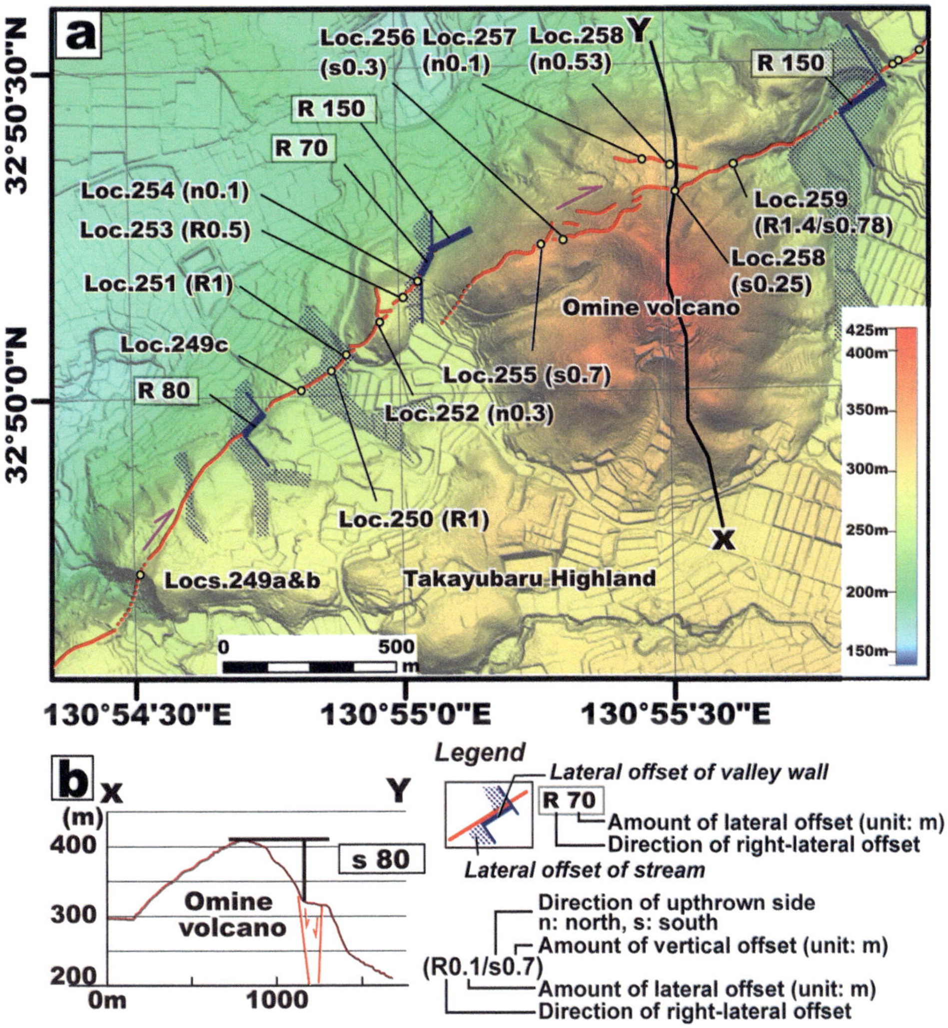

Fig. 10.59 a Detailed map of the surface ruptures and cumulative offset of streams and the outskirts of the Omine Volcano. See Figs. 10.4 and 10.5 for the location of the map. b Topographic profile across the surface ruptures and volcano body based on the DEM of the Geospatial Information Authority of Japan

Fig. 10.60 Oblique air photo of the two trenches at Komori. The small trench was excavated across the surface rupture by Iwasa et al. (2022) beside the already-dug big trench. Red arrows indicate the surface rupture of the 2016 earthquake

Fig. 10.61 Photograph of the elongated depression on the vegetable field at Loc. 249c before the trench excavation survey. White arrows indicate the direction of ground tilts toward the bottom of the depression. Modified after Iwasa et al. (2022)

Fig. 10.62 Photograph of two trenches across the surface rupture at Loc. 249c. On the south wall of the big trench, the boundary between the brown volcanic layer and black organic layer (white arrows) is tilted by the cumulative deformation of the fault

Fig. 10.63 Photograph of the deformed volcanic soil layers on the east wall of the small trench at Loc. 249c (Iwasa et al. 2022). Yellow and white arrows indicate the deformation during the 2016 earthquake and cumulative deformation, respectively

structure on the road (Fig. 10.88). There is a 400-m-long, 175-m-wide, and 25–30-m-high small isolated hill south of Tateno (Figs. 10.89 and 10.90). The distributed surface ruptures appeared along the northern margin of the hill as well as on the ridge, suggesting that the hill is a pressure ridge associated with right-lateral faulting. To the northeast, the rupture crossed a deep gorge of the Kurokawa River.

Inside the caldera, the surface rupture appears to have branched into three principal rupture zones (Fig. 10.7). The southern zone consisted of N80°E-trending surface ruptures that appeared for a distance of 1.8 km along the Nigorikawa River. In many places, narrow graben-like structures were formed between parallel faults (Figs. 10.91, 10.92, 10.93, 10.94, 10.95, 10.96, 10.97, 10.98, and 10.99). There was no

Fig. 10.64 Photograph of right-lateral offset (amount 100 cm) of the footpath of the vegetable field at Loc. 250. White and red arrows indicate the offset marker and surface rupture, respectively. There was no rupture on the field to the left due to the post-earthquake plowing by a tractor

Fig. 10.65 Photograph of right-lateral offset (amount 100 cm) of the sidewalk at Loc. 251. White and red arrows indicate the offset marker and surface rupture, respectively

Fig. 10.66 Photograph of north-up vertical displacement (amount 30 cm) on the hill slope at Loc. 252. Red arrows indicate the surface rupture

apparent vertical displacement across the grabens nor the lateral offsets. At Loc. 383h, two parallel traces appeared, and ~70 cm vertical separation was measured on both traces (Fig. 10.98). We also identified a preexisting ~17-m-wide and 4-m-deep graben on the volcanic foothill, and the co-seismic surface ruptures emerged within the graben.

Close to the central zone, a massive landslide occurred from the top of the caldera wall down to the Kurokawa River between Tateno and Kawayo (Figs. 10.100 and 10.101). Across the gorge to the northeast, the main surface ruptures of the central zone trended N65°E for a length of ~3 km with a left-stepping *en echelon* geometry across lava flow surfaces (Figs. 10.7 and 10.8). Around Kawayo, there were

Fig. 10.67 Photograph of right-lateral offset (amount 50 cm) of the fence at Loc. 253. White and red arrows indicate the offset marker and surface rupture, respectively

Fig. 10.68 Photograph of right-lateral offset of the dirt road at Loc. 254a. White and red arrows indicate the offset marker and surface rupture, respectively

Fig. 10.69 Photograph of the south-up vertical displacement at Loc. 254b. Red arrows indicate the surface rupture

two surface rupture traces: the main traces with predominant right-lateral offsets and secondary traces to the east with north-up vertical displacement (Fig. 10.102). Around Loc. 385 on the main trace, we measured the maximum right-lateral offset from 130 to 150 cm (Figs. 10.103, 10.104, 10.106, and 10.107). Here, a 10-m-wide graben structure was recognized over a length of ~100 m. A pit survey across the rupture at Loc. 385 showed that the volcanic soil layers overlying the lava flows were cumulatively displaced (Fig. 10.105). Near Loc 386a, a total of 150 cm vertical displacement was accommodated by several parallel ruptures (Fig. 10.108). There was another zone of rupture traces ~200 m to the south that passed through a dense residential area (Fig. 10.102), and many houses collapsed

Fig. 10.70 Photograph of the south-up vertical displacement (amount 70 cm) on a hill slope at Loc. 255

Fig. 10.71 Photograph of the north-up vertical displacement (amount 53 cm) on a hill slope at Loc. 258. Red arrows indicate the surface rupture

Fig. 10.72 Photograph of the south-up vertical displacement (amount 25 cm) on a hill slope at Loc. 259. Red arrows indicate surface rupture

because of the scattered ruptures (Figs. 10.109 and 10.110). Along the main trace, we could trace the ruptures nearly continuously between Locs. 387 and 392. The right-lateral displacement was 60 cm at Loc. 387 and decreased to the northeast (Figs. 10.111, 10.112, 10.113 and 10.114). Toda et al. (2019) conducted a trench survey at Locs. 388 and 389, and identified four paleoseismic events after the deposition of the 7.3 ka Kikai-Akahoya tephra layer including the 2016 event. About ~100 m north of Loc. 389a, there was another ~300-m-long right-lateral rupture trace, and N80° W-trending conjugate ruptures emerged for a length of ~80 m to bridge the two parallel surface rupture traces (Figs. 10.115 and 10.116). At Loc. 393 on the northern trace of the two parallel ruptures, the right-lateral offset was 55 cm (Fig. 10.117). Further north at Loc. 394, an E-W-trending another conjugate trace appeared over ~50 m, where 8 cm left-lateral offset was observed (Fig. 10.118). To the east, another trace appeared from west of Loc. 395 to Loc. 397 for ~300 m with a right-lateral displacement of ~30 cm (Figs. 10.119 and 10.120).

Fig. 10.73 Photograph of right-lateral offset (amount 140 cm) of the edge of the footpath with south-up vertical displacement (amount 78 cm) at Loc. 260. White and red arrows indicate the offset marker and surface rupture, respectively

Fig. 10.74 Photograph of surface rupture cutting through the western bank of the Ookirihata reservoir at Loc. 260a. White and red arrows indicate the offset marker and surface rupture, respectively

Fig. 10.75 Photograph of right-lateral offset (amount ~ 145 cm) of the orange-colored center line on the road at Loc. 320. White and red arrows indicate the offset marker and surface ruptures, respectively. Yellow arrows indicate minor conjugate rupture with left-lateral offset. The right-lateral offset of the guardrail at Loc. 321 is also visible on the right

In the northern zone, the N70°E-trending ruptures extended across the campus of Kyushu Tokai University from Loc. 398, with a N35°W-trending and ~160-m-long widely distributed ruptures cutting across the university athletic field to the south (Loc. 397a; Fig. 10.121). The main fault zone consisted of two straight traces, but the southern trace was discontinuous and unclear. The northern trace extended beyond the university campus for 2.5 km and nearly continuous to its northeastern end (Figs. 10.7, 10.8, 10.122, 10.123, 10.124, 10.125, 10.126, 10.127, 10.128, 10.129, 10.130, 10.131, and 10.132). The average right-lateral offset on this section was ~50 cm with the

Fig. 10.76 Photograph of right-lateral offset (amount 90 cm) of the boundary between the grass and gravel surfaces with north-up vertical deformation (amount 20 cm) at Loc. 327. White and red arrows indicate the offset marker and surface rupture, respectively

Fig. 10.77 Photograph of right-lateral offset (amount 35 cm) of the edge of the paved road at Loc. 328. White and red arrows indicate the offset marker and surface rupture, respectively

Fig. 10.78 Photograph of left-stepping surface ruptures (red arrows) at Loc. 329

Fig. 10.79 Photograph of surface rupture (red arrows) with north-up vertical displacement (amount 20 cm) at Loc. 331

Fig. 10.80 Photograph of right-lateral offset (amount 160 cm) of the road's edge at Loc. 334. White and red arrows indicate the offset marker and surface rupture, respectively

Fig. 10.81 Photograph of right-lateral offset (amount 165 cm) of the road's edge at Loc. 336. White and red arrows indicate the offset marker and surface rupture, respectively

Fig. 10.82 Photograph of right-lateral offset (amount 120 cm) of the faint white center line on the road at Loc. 339. White and red arrows indicate the offset marker and surface rupture, respectively

Fig. 10.83 Photograph of surface rupture with north-up vertical displacement (amount 20 cm) at Loc. 342. Red arrows indicate the surface rupture

Fig. 10.84 Photograph of surface rupture with north-up vertical deformation (amount 65 cm) at Loc. 380a. The rupture was also associated with ~50-cm right-lateral displacement. White and red arrows indicate the offset marker and surface rupture, respectively

Fig. 10.85 Photograph of north-up vertical deformation (amount 35 cm) along a small valley at Loc. 380. Red arrows indicate the surface rupture

Fig. 10.86 **a** Photograph of a fault outcrop at Loc. 381 on the right bank of the Shirakawa River. A zone between white and red dotted lines is a fresh and unweathered part of the fault plane, suggesting that it was exposed by fault movement during the 2016 earthquake. The right-lateral offset was ~140 cm. **b** Close-up photograph of the near-vertical fault plane with a ~30-cm-wide shear zone

Fig. 10.87 Photograph of right-lateral offset of the train rail at Loc. 381a. White and red arrows indicate the offset marker and surface rupture, respectively

Fig. 10.88 Photograph of left-stepping surface ruptures with a right-lateral offset of the white line along the road at Loc. 381b. White and red arrows indicate the offset marker and surface ruptures, respectively

Fig. 10.89 Photograph of the isolated hill or pressure ridge at Tateno around Loc. 382. The surface rupture appeared along the northern margin of the ridge as well as on the ridge. Red arrows indicate the surface rupture

Fig. 10.90 Photograph of the isolated hill the pressure ridge at Tateno around Loc. 382. The surface rupture appeared along the northern margin of the ridge as well as on the ridge. Red arrows indicate the surface rupture

Fig. 10.91 Photograph of surface rupture that appeared as open cracks at Loc. 383a. Red arrows indicate the surface rupture

Fig. 10.92 Photograph of the surface ruptures (open cracks) at Loc. 383b. Red arrows indicate the surface ruptures. These parallel ruptures formed a graben

Fig. 10.93 Photograph of the surface ruptures that appeared as a 1.4-m-wide open crack at Loc. 383c

Fig. 10.94 Photograph of the surface rupture at Loc. 383d. Red arrows indicate the fault trace with south-down vertical deformation

Fig. 10.95 Photograph of the surface rupture that appeared as a 1-m-wide open crack at Loc. 383e

Fig. 10.96 Photograph of the surface rupture that appeared as a 0.5-m-wide open crack at Loc. 383f. Red arrows indicate the surface rupture

Fig. 10.97 Photograph of the surface rupture with south-up vertical displacement (amount 30 cm) at Loc. 383g. Red arrows indicate the surface rupture

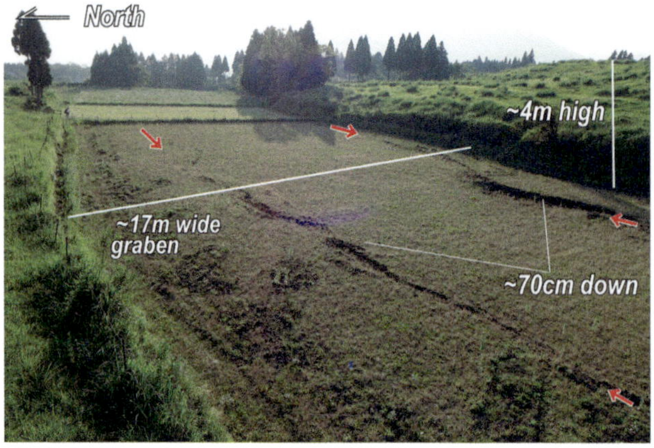

Fig. 10.98 Photograph of two parallel surface ruptures within the preexisting ~17-m-wide and 4-m-deep graben on the volcanic fan surface at Loc. 383h. The amounts of vertical displacement along both ruptures were ~70 cm. Red arrows indicate the surface ruptures

Fig. 10.99 Photograph of three parallel surface ruptures forming a graben at Loc. 383i. Red arrows indicate the surface ruptures

Fig. 10.100 Oblique air photo south of Minami-Aso Village. Red arrows indicate the approximate location of the surface ruptures. The photo was taken by Akihiro Murata, Professor Emeritus of Tokushima University

Fig. 10.101 Oblique air photo around Kawayo, Minami-Aso Village. Red arrows indicate the approximate location of the surface ruptures. The photo was taken by Akihiro Murata, Professor Emeritus of Tokushima University

Fig. 10.102 Detailed map of the surface ruptures around Kawayo and Kurokawa, Minami-Aso Village. Red and orange lines indicate the field-checked and site-indistinct traces, respectively. The location of the map is shown in Fig. 10.7

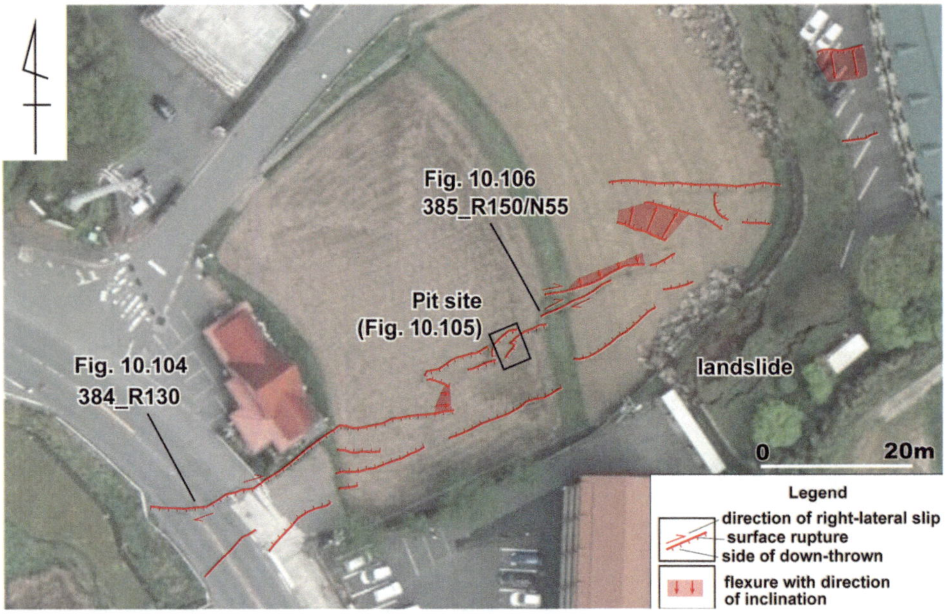

Fig. 10.103 Detailed map of the surface ruptures around Loc. 385 at Kawayo, Minami-Aso Village. Parallel ruptures formed a narrow ~100-m-long graben. The location of the map is shown in Fig. 10.102

Fig. 10.104 Photograph of right-lateral offset (amount 130 cm) of the centerline on the road at Loc. 384. White and red arrows indicate the offset marker and surface rupture, respectively

Fig. 10.105 West wall of the pit across the surface rupture near Loc. 385. Layers exposed on the wall had been deformed and displaced more significantly than by the 2016 earthquake. Red arrows indicate the fault plane. The location of the pit is shown in Fig. 10.103

Fig. 10.106 Photograph of right-lateral offset (amount 150 cm) of the footpath with north-up deformation (amount 55 cm) at Loc. 385. White and red arrows indicate the offset marker and surface rupture, respectively

Fig. 10.107 Photograph of right-lateral offset of the foundation of a damaged apartment at Loc. 385a. White and red arrows indicate the offset marker and surface rupture, respectively

Fig. 10.108 Photograph of parallel ruptures with north-up displacement (total amount 150 cm) at Loc. 386a. Red arrows indicate the surface ruptures

Fig. 10.109 Photograph of a series of parallel ruptures with north-up displacement at Loc. 386b. Red arrows indicate the surface ruptures

Fig. 10.110 Photograph of a series of parallel ruptures with north-up displacement at Loc. 386c. Red arrows indicate the surface ruptures

Fig. 10.111 Photograph of right-lateral offset of ridges of vegetable field (amount 15 cm) at Loc. 388. White and red arrows indicate the offset marker and surface rupture, respectively

Fig. 10.112 Photograph of the surface rupture at Loc. 389a. Red arrows indicate the surface rupture

Fig. 10.113 Photograph of right-lateral offset of the road's edge (amount 50 cm) at Loc. 390. White and red arrows indicate the offset marker and surface rupture, respectively

Fig. 10.114 Photograph of the surface rupture with south-up vertical displacement (amount 10 cm) at Loc. 392. Red arrows indicate the surface rupture

Fig. 10.115 Photograph of the N80°W-trending conjugate rupture trace on the paddy field around Loc. 392a. Red arrows indicate the surface rupture

Fig. 10.116 Photograph of the N80°W-trending conjugate rupture trace that bridged the two NE-trending parallel surface ruptures across the paddy field around Loc. 392a. Red arrows indicate the surface ruptures

Fig. 10.117 Photograph of right-lateral offset of a grass line on the paddy field (amount 55 cm) at Loc. 393. White and red arrows indicate the offset marker and surface rupture, respectively

Fig. 10.118 Photograph of the conjugate left-lateral offset (amount 8 cm) of the white line on the parking area around Loc. 394. White and red arrows indicate the offset marker and surface rupture, respectively

Fig. 10.119 Photograph of right-lateral offset of the straight line of stubble on the paddy field (amount 34 cm) with south-up vertical displacement (amount 18 cm) at Loc. 395. White and red arrows indicate the offset marker and surface rupture, respectively

Fig. 10.120 Photograph of right-lateral offset of the ditch on the parking area (amount 30 cm) at Loc. 396. White and red arrows indicate the offset marker and surface rupture, respectively

Fig. 10.121 Photograph of the surface rupture that cut across the athletic field of Kyushu Tokai University around Loc. 397a. Red arrows indicate the surface rupture

Fig. 10.122 Photograph of right-lateral offset of the ridges on the vegetable field (amount 10 cm) at Loc. 403. White and red arrows indicate the offset marker and surface rupture, respectively

Fig. 10.123 Photograph of right-lateral offset of the edge of asphalt-covered cowshed site (amount 45 cm) with north-up vertical displacement (amount 20 cm) at Loc. 406. White and red arrows indicate the offset marker and surface rupture, respectively

Fig. 10.124 Photograph of north-up vertical displacement (amount 35 cm) with a right-lateral offset of the road's edge (amount 12 cm) at Loc. 407. White and red arrows indicate the offset marker and surface ruptures, respectively

Fig. 10.125 Photograph of right-lateral offset of the stonework in the flower bed (amount 50 cm) with north-up vertical displacement (amount 25 cm) at Loc. 410. White and red arrows indicate the offset marker and surface ruptures, respectively

Fig. 10.126 Photograph of north-up vertical deformation (amount 55 cm) at Loc. 411. Red arrows indicate the surface rupture

Fig. 10.127 Photograph of right-lateral offset of the house and paved road (amount 110 cm) at Loc. 412. Red arrows indicate the surface rupture

Fig. 10.128 Photograph of right-lateral offset of the boundary of different lawn mowing (amount 45 cm) with north-up vertical displacement (amount 10 cm) at Loc. 413. White and red arrows indicate the offset marker and surface rupture, respectively

Fig. 10.129 Photograph of surface rupture at Loc. 413. Red arrows indicate the surface rupture

Fig. 10.130 Photograph of right-lateral offset of the road's edge (amount 30 cm) at Loc. 415. White and red arrows indicate the offset marker and surface rupture, respectively

Fig. 10.131 Photograph of the surface rupture with a right-lateral offset of 45 cm at Loc. 417. Red arrows indicate the surface rupture

Fig. 10.132 Photograph of right-lateral offset of the road's edge (amount 10 cm) at Loc. 419. White and red arrows indicate the offset marker and surface rupture, respectively

maximum amount of 110 cm at Loc. 412 (Fig. 10.127). The vertical displacement was north-up, and its amount ranged from 10 to 55 cm.

Acknowledgements We would like to thank Akihiro Murata, Professor Emeritus, Tokushima University for providing us with spectacular air photos. This study was supported by the JSPS grants 18H03601.

References

Ikeda Y, Chida N, Nakata T, Kaneda H, Tajikara M, Takazawa N (2001) 1:25,000 Active Fault Map, Kumamoto. Geospatial Information Authority of Japan

Ishimura D (2019) Co-seismic vertical displacement associated with the 2016 Kumamoto earthquake (Mw7.0) and activity of the Futagawa fault around Futa, Nishihara Village, Kumamoto Perfecture. Active Fault Res 50:33–44

Iwasa Y, Kumahara Y, Goto H, Nakata T (2020) Detailed mapping of surface ruptures associated with the 2016 Kumamoto earthquake and faulting history of the conjugated fault in Dozon, Mashiki Town, Kumamoto Prefecture. Active Fault Res 52:1–8

Iwasa Y, Kumahara Y, Goto H, Ishimura D, Hosoya T (2022) Faulting history of the Futagawa fault zone based on trenching survey at Komori, Nishihara Village, Kumamoto Prefecture. Active Fault Res 56:47–58

Machida H, Arai F (2003) Atlas of Tephra in and around Japan. University of Tokyo Press, Tokyo

Matsumoto A, Uto K, Ono K, Watanabe K (1991) K-Ar age determinations for Aso volcanic rocks: concordance with volcanostratigraphy and application to pyroclastic flows. Program Abstracts Volcanol Soc Japan 2:73. https://doi.org/10.18940/vsj.1991.2.0_73

Miyoshi M, Shinmura T, Sumino H, Sano T, Miyabuchi Y, Mori Y, Inakura H, Furukawa K, Uno K, Hasenaka T, Nagao K, Arakawa Y, Yamamoto J (2013) Lateral magma intrusion from a caldera-forming magma chamber: constraints from geochronology and geochemistry of volcanic products from lateral cones around the Aso caldera, SW Japan. Chem Geol 352:202–210. https://doi.org/10.1016/j.chemgeo.2013.06.003

Smith VC, Staff RA, Blockley SPE, Bronk Ramsey C, Nakagawa T, Mark DF, Takemura K, Danhara T, Suigetsu 2006 Project Members (2013) Identification and correlation of visible tephras in the Lake Suigetsu SG06 sedimentary archive, Japan: chronostratigraphic markers for synchronising of east Asian/west Pacific palaeoclimatic records across the last 150 ka. Quat Sci Rev 67:121–137. https://doi.org/10.1016/j.quascirev.2013.01.026

Toda S, Torii M, Okuno M, Konno A, Ono H, Takahashi N (2019) Evidence for Holocene paleoseismic events on the 2016 Kumamoto earthquake rupture zone within the Aso caldera: a trench excavation survey at Kurokawa, the town of Minami-Aso, southwest Japan. Active Fault Res 51:13–25

Tsutsumi H, Toda S, Goto H, Kumahara K, Ishimura D, Takahashi N, Taniguchi K, Omata M, Kohriya Y, Gomi M, Asano K, Iwata T (2018) Paleoseismic trenching across the surface rupture of the 2016 Kumamoto earthquake at Jichu, Mashiki Town, Kumamoto Perfecture. Active Fault Res 49:31–39

Watanabe K, Ono K (1969) Geology of the vicinity of Omine on the western flank of the Aso caldera. J Geol Soc Japan 75:365–374. https://doi.org/10.5575/geosoc.75.365

Surface Ruptures in the Downtown of Kumamoto City

11

Hideaki Goto, Shinji Toda, Hiroyuki Tsutsumi, and Yasuhiro Kumahara

Abstract

Surface ruptures were identified along the NW-trending flexure scarps called the Suizenji fault zone in the downtown of Kumamoto City, based on the interpretation of the InSAR image. In this chapter, we describe results of field surveys conducted to examine whether surface deformation such as open cracks, tilts, and shortenings appeared on the surface along the fault zone. The vertical deformation associated with the 2016 Kumamoto earthquake was quite small (<~0.1 m) compared to the empirically calculated coseismic slip (~0.5 m) based on the fault length. It is suggested that these slips on this fault zone were triggered by the Kumamoto earthquake.

Keywords

Surface deformation • Triggered slip • InSAR • Suizenji fault zone

Goto et al. (2017) interpreted topographic stereo images (anaglyph) made from the 5-m-grid digital elevation model of the Geospatial Information Authority of Japan (GSI), which itself was generated from light detection and ranging (LiDAR) data. They identified several NW-trending flexure scarps down to the southwest that deform several SW-sloping river terraces formed after the deposition of the pyroclastic flow deposits related to the latest massive eruption of the Aso Volcano (86.8–87.3 ka; Aoki 2008) (Figs. 11.1 and 11.2). Goto et al. (2017) named these 5.4-km-long scarps that cut across the downtown of Kumamoto as the Suizenji fault zone.

Interpretation of the InSAR images of Geospatial Information Authority of Japan (2016) and Fujiwara et al. (2016) led them to recognize ~10 cm coseismic deformation along the Akitsukawa flexure zone and Suizenji fault zone (Fig. 11.3). Motivated by the image, they conducted field surveys to reveal whether surface deformation such as open cracks, tilts, and shortenings appeared on the surface along these zones. In this chapter, we describe more detailed results of the field surveys of Goto et al. (2017) with larger scale location maps and higher resolution field photographs.

In many places along F1 of the Suizenji fault zone, a series of continuous open cracks were observed on the artificial features like asphalt roads and retaining walls (Fig. 11.4). At Loc. K01, there is a coseismic open crack at the base of the F1 fault scarp (Fig. 11.5). The Kumamoto Technical High School swimming pool (Loc. K02) is located on a west-facing flexure scarp of F1. The trace of the pre-earthquake water line was tilted to the west by ~ 8 cm

Supplementary Information
The online version contains supplementary material available at (https://doi.org/10.1007/978-981-19-1150-7_11).
Note
The supplementary material of this book contains geographical information system (GIS) and spreadsheet files of the surface ruptures associated with the mainshock of the 2016 Kumamoto earthquake sequence

H. Goto (✉)
Graduate School of Humanities and Social Sciences, Hiroshima University, 1-2-3, Kagamiyama, Higashi-Hiroshima, Hiroshima, 739-8522, Japan
e-mail: hgoto@hiroshima-u.ac.jp

S. Toda
International Research Institute of Disaster Science (IRIDeS), Tohoku University, 468-1, Aoba, Aramaki, Aoba-ku, Sendai, Miyagi, 980-0845, Japan
e-mail: toda@irides.tohoku.ac.jp

H. Tsutsumi
Department of Environmental Systems Science, Faculty of Science and Engineering, Doshisha University, 1-3, Tatara-Miyakodani, Kyotanabe, Kyoto, 610-0394, Japan
e-mail: htsutsum@mail.doshisha.ac.jp

Y. Kumahara
Graduate School of Humanities and Social Sciences, Hiroshima University, 1-1-1, Kagamiyama, Higashi-Hiroshima, Hiroshima, 739-8524, Japan
e-mail: kumakuma@hiroshima-u.ac.jp

Y. Kumahara et al. (eds.), *Surface Ruptures Associated with the 2016 Kumamoto Earthquake Sequence in Southwest Japan*, Advances in Geological Science, https://doi.org/10.1007/978-981-19-1150-7_11

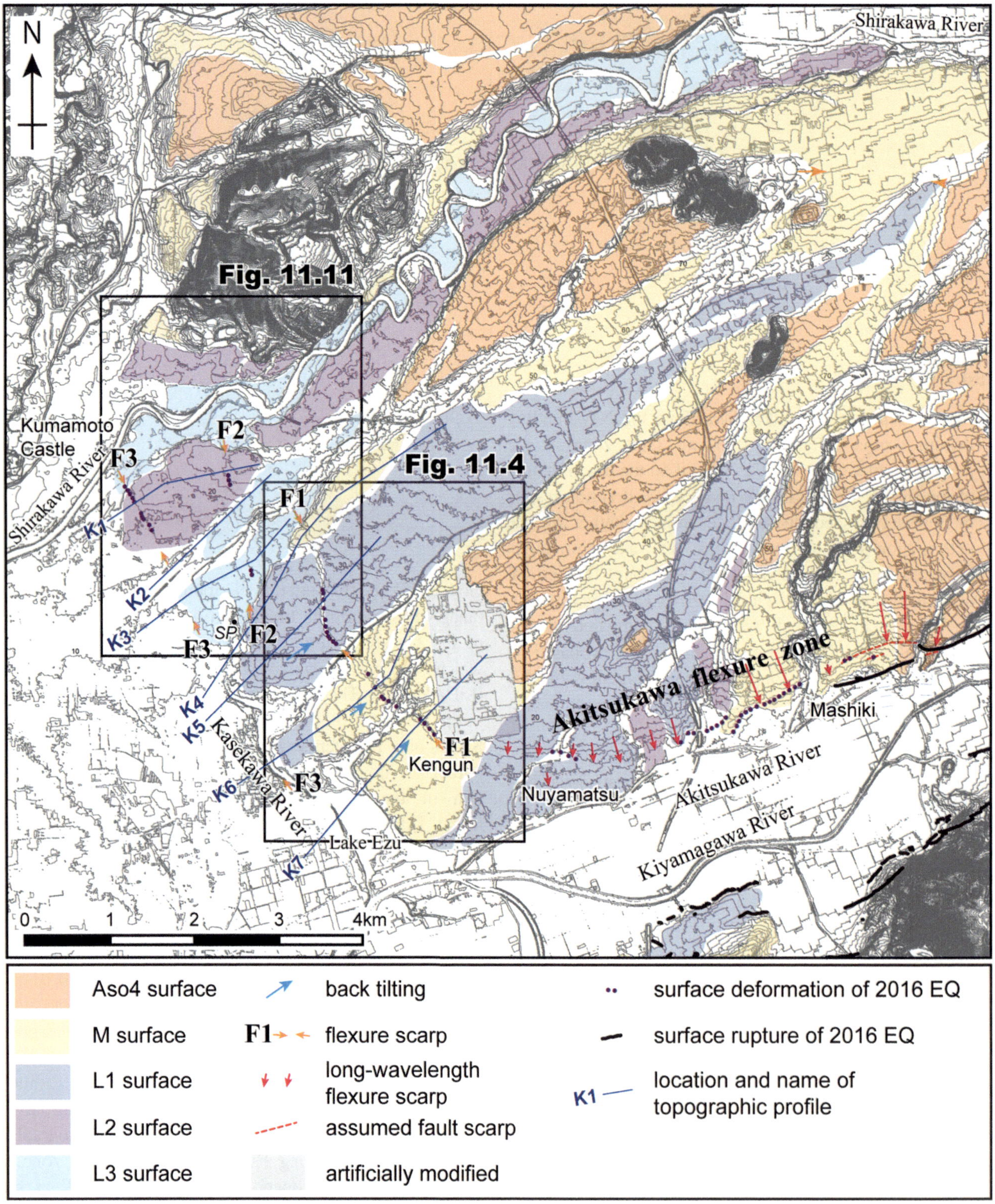

Fig. 11.1 Geomorphic and active fault map in Mashiki and the downtown of Kumamoto City. The contour map is based on the 5-m-grid digital elevation model of the Fundamental Geospatial Data issued by the GSI. SP: Suizenji Jojuen Park (natural spring pond). After Goto et al. (2017)

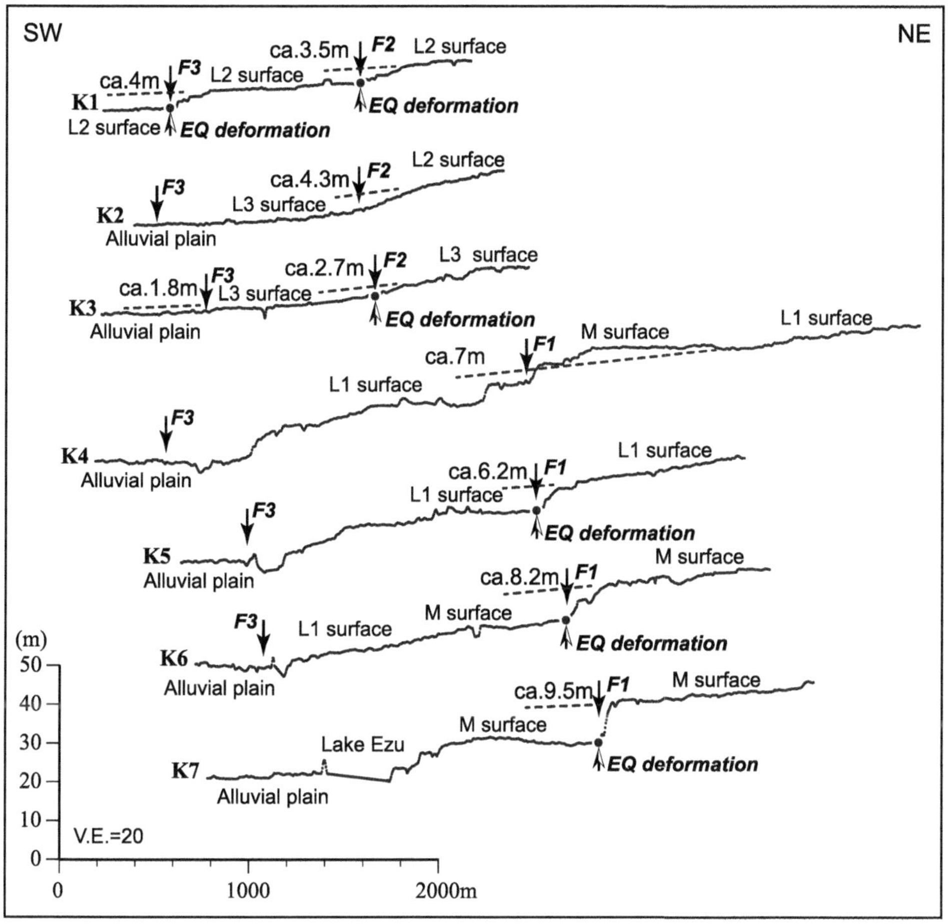

Fig. 11.2 Geomorphic profiles across the ruptures of the 2016 Kumamoto earthquake. Locations of the profiles are shown in Fig. 11.1

due to coseismic deformation (Fig. 11.6). These deformational features indicate that the flexural scarps grew during the earthquake.

At Loc. K03, open cracks appeared on the road south of the Kumamoto Technical High School in the middle of the fault scarp (Fig. 11.7). These ruptures continued southeast to the retaining walls and a high-rise apartment barely visible on the right of the photograph of Fig. 11.7.

We identified the buckling of asphalt at the base of the scarp at Loc. K04 south of the high-rise apartment (Fig. 11.8). The tombstones of the cemetery at Loc. K05 were heavily damaged near the ruptures that appeared in the middle of the fault scarp (Fig. 11.9). The residents told us that these deformational features along F1 appeared at midnight of April 16 during the mainshock event.

The ruptures were indistinct at the left bank of the Kengun River for ~ 200 m, while the ruptures were recognized from south of the Kengun Shrine to the Kengun tram station. We identified open cracks on the roads across the fault scarp from Locs. K06 to K09. At Loc. K08, cracks appeared at the base of the fault scarp and the retailing wall to the northwest (Fig. 11.10).

Fault-related deformation, such as open cracks, was found at only a few places along F2 of the Suizenji fault

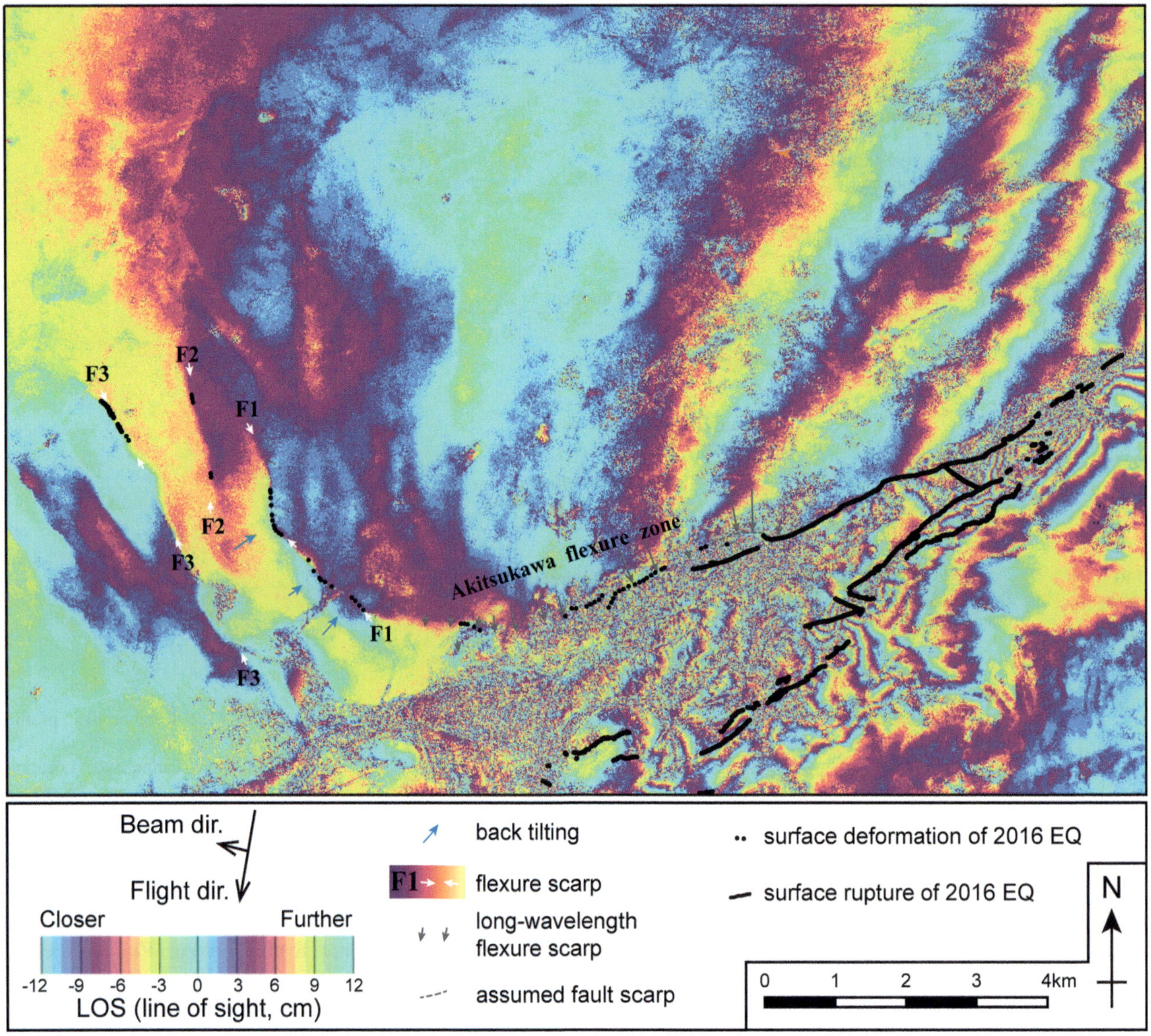

Fig. 11.3 Distribution of active faults and surface deformation associated with the 2016 Kumamoto earthquake overlaid on the InSAR deformation map, descending interferograms computed from ALOS-2/PALSAR-2 images of March 7, 2016 and April 18, 2016 (Geospatial Information Authority of Japan, 2016). After Goto et al. (2017)

zone, suggesting that the displacement amount was smaller than that along F1.

Along the northern part of F3 of the Suizenji fault zone (Fig. 11.11), continuous surface deformation, including right-stepping *en echelon* open cracks and left-lateral offset of the concrete block wall, was observed with an amount of ~ 0.5 cm at Loc. K10 (Fig. 11.12). Houses and concrete buildings standing between the visible deformation features

Fig. 11.4 Detailed surface rupture map around Kengun of the downtown Kumamoto City. The background is a half-transparent topographic map of Geospatial Information Authority of Japan (GSI) that is underlain by a slope map produced from 5-m-grid digital elevation models (DEMs) of GSI. The location of the map is shown in Fig. 11.1

Fig. 11.5 Photograph of surface rupture at Loc. K01. Looking to the south. White arrows indicate a small open crack as an indicator of the earthquake rupture. After Goto et al. (2017)

Fig. 11.6 Photograph of tilting deformation of a swimming pool during the earthquake at Loc. K02. Looking to the west. White arrows indicate the water level of the pool before the earthquake. After Goto et al. (2017)

Fig. 11.7 Photograph of surface rupture at Loc. K03. Looking to the east. White arrows indicate small open cracks as an indicator of rupture. After Goto et al. (2017)

Fig. 11.8 Photograph of the surface rupture at Loc. K04. Looking to the east. Red and white arrows indicate small open cracks in the middle of the fault scarp and shortening breaks at the base of the fault scarp, respectively, as an indicator of rupture. After Goto et al. (2017)

Fig. 11.9 Photograph of heavily damaged tombstones at Loc. K05. Looking to the northwest. White arrows indicate a ground rupture

Fig. 11.10 Photograph of the surface rupture at Loc. K08. Looking to the north. White arrows indicate small open cracks as an indicator of rupture. After Goto et al. (2017)

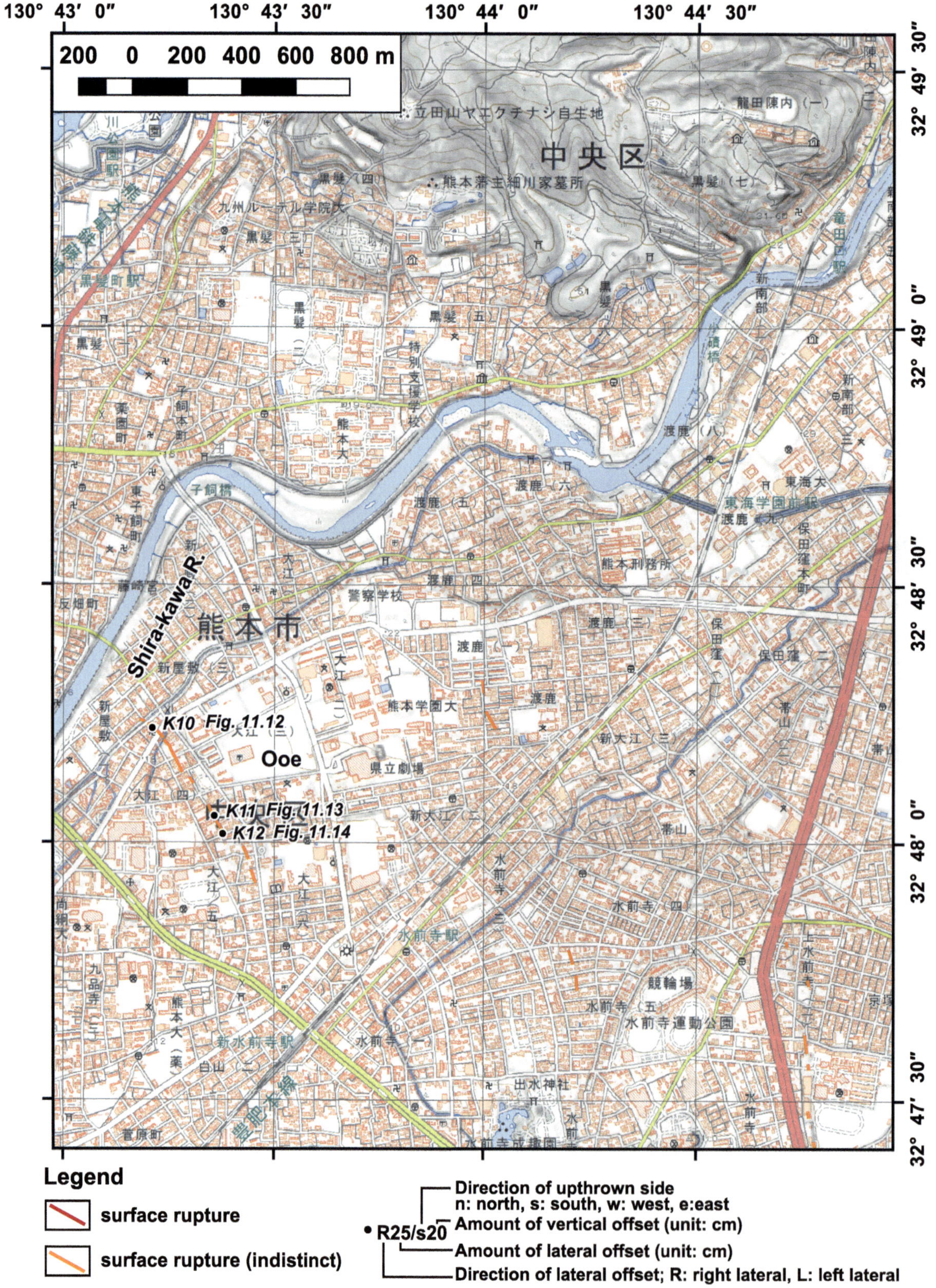

Fig. 11.11 Detailed surface rupture map around the downtown Kumamoto City. The background is a half-transparent topographic map of Geospatial Information Authority of Japan (GSI) that is underlain by a slope map produced from 5-m-grid digital elevation models (DEMs) of GSI. The location of the map is shown in Fig. 11.1

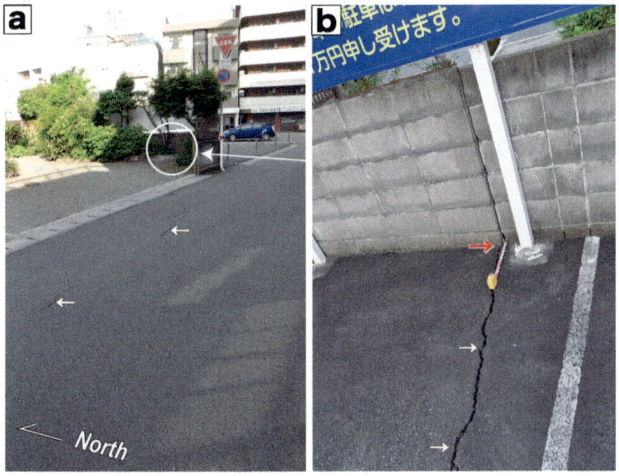

Fig. 11.12 Photographs of surface rupture at Loc. K10. **a** Looking to the south. **b** Looking to the north at the other side of the block wall indicated by a white circle on the panel **a**. Red and white arrows indicate the left-lateral offset of block walls and open cracks, respectively, as indicators of rupture. After Goto et al. (2017)

Fig. 11.13 Photograph of surface rupture at Loc. K11. Looking to the north. White arrows indicate small open cracks as an indicator of rupture. After Goto et al. (2017)

like open cracks at Loc. K11 (Fig. 11.13) were heavily damaged, implying that the structural basements might have been deformed by faulting as well. The doors of the concrete building beside the right-stepping *en echelon* open cracks became unfunctional due to the deformation of the base and frame at Loc. K12 (Fig. 11.14). The residents identified the disruption on April 16 morning after the mainshock of the 2016 Kumamoto earthquake.

As we described above, the amount of vertical deformation associated with the 2016 Kumamoto earthquake was quite small ($<\sim$ 0.1 m) compared to the empirically calculated coseismic slip ($\sim$0.5 m) based on the fault length. It is

Fig. 11.14 Photograph of right-stepping *en echelon* open cracks at Loc. K12. Looking to the south. Red arrow denotes location of the crack under the unfunctional door. White arrows indicate a ground rupture. After Goto et al. (2017)

suggested that these slips on this fault zone were triggered by the Kumamoto earthquake.

Acknowledgements This work was partly supported by MEXT KAKENHI Grant 16H06298 and 16K01221. We thank the staff of Kumamoto Technical High School, who allowed us to conduct our fieldwork inside the school property.

References

Aoki K (2008) Revised age and distribution of ca. 87 ka Aso-4 tephra based on new evidence from the northwest Pacific Ocean. Quatern Int 178:100–118

Fujiwara S, Yarai H, Kobayashi T, Morishita Y, Nakano T, Miyahara B, Nakai H, Miura Y, Ueshiba H, Kakiage Y, Une H (2016) Small-displacement linear surface ruptures of the 2016 Kumamoto earthquake sequence detected by ALOS-2 SAR interferometry. Earth Planets Space 68:160. https://doi.org/10.1186/s40623-016-0534-x

Geospatial Information Authority of Japan (2016) Detection of crustal movement by ALOS-2 InSAR. http://www.gsi.go.jp/BOUSAI/H27-kumamoto-earthquake-index.html#3. Accessed 7 July 2016

Goto H, Tsutsumi H, Toda S, Kumahara Y (2017) Geomorphic features of surface ruptures associated with the 2016 Kumamoto earthquake in and around the downtown of Kumamoto City, and implications on triggered slip along active faults. Earth Planets Space 69:26 https://doi.org/10.1186/s40623-017-0603-9

Surface Ruptures and Tectonic Geomorphology Along and Around the Idenokuchi Fault

12

Heitaro Kaneda, Shinji Toda, Daisuke Ishimura, Yasuhiro Kumahara, Hideaki Goto, Shinsuke Okada, and Motoya Kobayashi

Abstract

During the mainshock of the 2016 Kumamoto earthquake sequence, the Idenokuchi fault, a normal fault that runs to the southeast of and subparallel to the right-lateral Futagawa fault, also ruptured the surface, suggesting slip-partitioned surface ruptures. The main surface rupture extended for a distance of ~10 km with the maximum vertical slip as large as ~1.9 m down to the northwest. However, the main rupture was accompanied with several traces of antithetic surface faulting to the northwest, with the maximum vertical slip of ~0.8 m down to the southeast. Almost all the surface ruptures along and around the Idenokuchi fault, including the antithetic traces, were associated with preexisting geomorphic signals of earlier surface faulting, although the northern half of the Idenokuchi fault had been largely unmapped before 2016. Careful examination of high-resolution LiDAR DEMs with the concept and possibility of slip partitioning and antithetic faulting would be a key for more complete active fault mapping.

Keywords

2016 Kumamoto earthquake sequence • Idenokuchi fault • Slip partitioning • Normal fault • LiDAR

Supplementary Information
The online version contains supplementary material available at (https://doi.org/10.1007/978-981-19-1150-7_12).
Note
The supplementary material of this book contains geographical information system (GIS) and spreadsheet files of the surface ruptures associated with the mainshock of the 2016 Kumamoto earthquake sequence

H. Kaneda (✉)
Department of Civil and Environmental Engineering, Chuo University, 1-13-27, Kasuga, Bunkyo-ku, Tokyo, 112-8551, Japan
e-mail: kaneda.00c@g.chuo-u.ac.jp

S. Toda
International Research Institute of Disaster Science (IRIDeS), Tohoku University, 468-1, Aoba, Aramaki, Aoba-ku, Sendai, Miyagi, 980-0845, Japan
e-mail: toda@irides.tohoku.ac.jp

D. Ishimura
Department of Geography, Tokyo Metropolitan University, 1-1, Minami-Osawa, Hachioji, Tokyo, 192-0397, Japan
e-mail: ishimura@tmu.ac.jp

Y. Kumahara
Graduate School of Humanities and Social Sciences, Hiroshima University, 1-1-1, Kagamiyama, Higashi-Hiroshima, Hiroshima, 739-8524, Japan
e-mail: kumakuma@hiroshima-u.ac.jp

H. Goto
Graduate School of Humanities and Social Sciences, Hiroshima University, 1-2-3, Kagamiyama, Higashi-Hiroshima, Hiroshima, 739-8522, Japan
e-mail: hgoto@hiroshima-u.ac.jp

S. Okada
Faculty of Science and Engineering, Iwate University, 4-3-5, Ueda, Morioka, Iwate, 020-8551, Japan
e-mail: sokada@iwate-u.ac.jp

M. Kobayashi
Keiyo Gas Co., Ltd., 2-8-8, Ichikawa-minami, Ichikawa, Chiba, 272-8580, Japan

12.1 Introduction

One of the interesting features of the 2016 Kumamoto earthquake was slip-partitioned surface ruptures (Toda et al. 2016). Approximately 1.2–2.0 km to the southeast of the extensive surface ruptures along the right-lateral Futagawa fault, a NE-trending normal fault called the Idenokuchi fault (Research Group for Active Tectonics in Kyushu 1989) also ruptured the surface for a distance of ~10 km (Fig. 6.1). The amount of vertical slip on the Idenokuchi fault was as large as ~1.9 m, comparable to the maximum dextral slip on the Futagawa fault (2.25 m; see Fig. 10.29). General features of the surface ruptures along the Idenokuchi fault

Y. Kumahara et al. (eds.), *Surface Ruptures Associated with the 2016 Kumamoto Earthquake Sequence in Southwest Japan*, Advances in Geological Science, https://doi.org/10.1007/978-981-19-1150-7_12

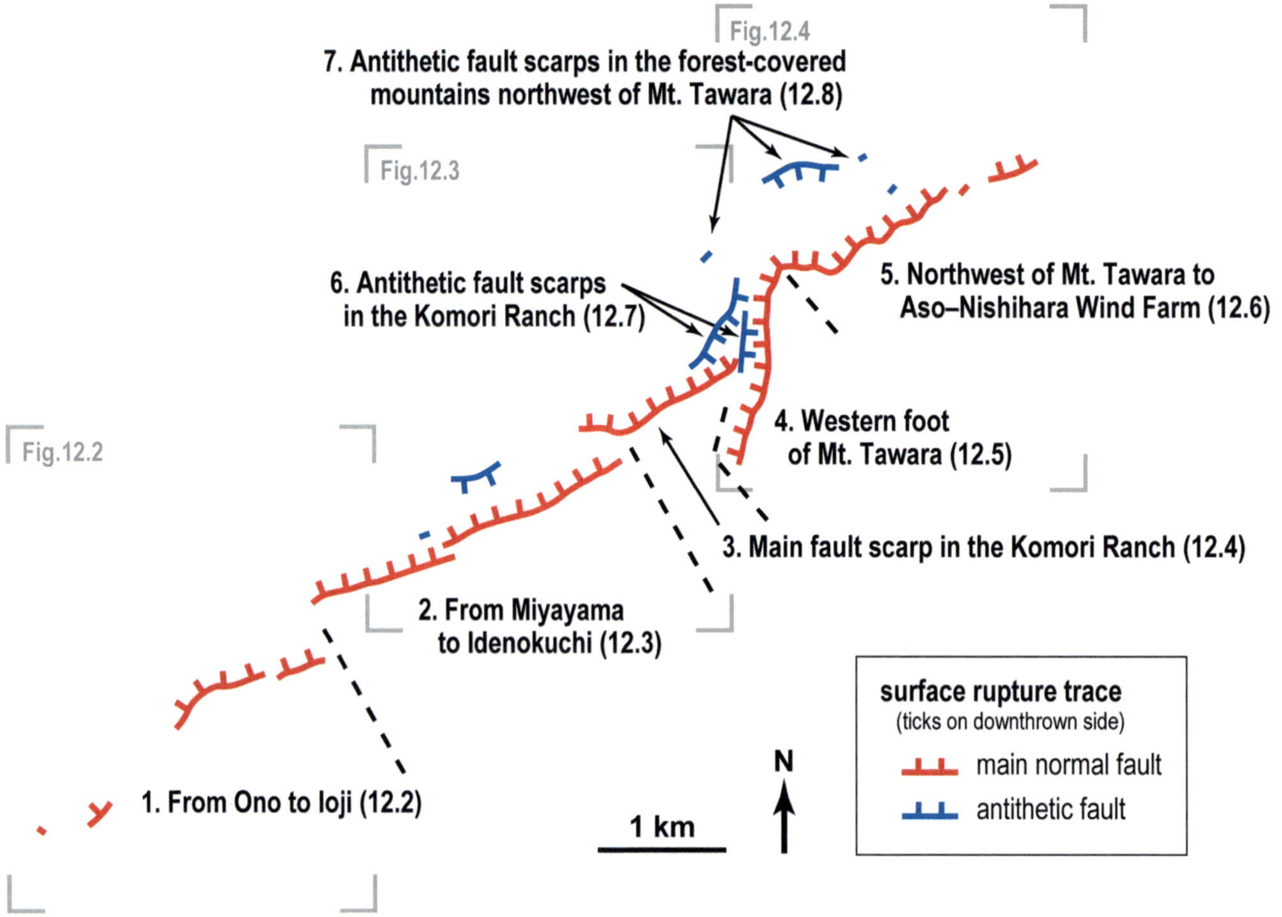

Fig. 12.1 Subsections of the surface ruptures along and around the Idenokuchi fault

were summarized by Toda et al. (2016), and here we describe much more details of those surface ruptures as well as their tectonic–geomorphic expressions, with the addition of some new information and discoveries after the publication of Toda et al. (2016).

The description of the surface ruptures is subdivided into seven subsections as follows (Fig. 12.1):

Subsection 1: From Ono to Ioji (12.2)
Subsection 2: From Miyayama to Idenokuchi (12.3)
Subsection 3: Main fault scarp in the Komori Ranch (12.4)
Subsection 4: Western foot of Mt. Tawara (12.5)
Subsection 5: Northwest of Mt. Tawara to Aso–Nishihara Wind Farm (12.6)
Subsection 6: Antithetic fault scarps in the Komori Ranch (12.7)
Subsection 7: Antithetic fault scarps in the forest-covered mountains northwest of Mt. Tawara (12.8).

More detailed surface rupture traces are shown in Figs. 12.2, 12.3, and 12.4. For each subsection, we describe details of the 2016 surface ruptures, followed by a short description of tectonic geomorphology along that subsection.

12.2 From Ono to Ioji

12.2.1 Surface Ruptures

The south-westernmost evidence of surface faulting along the Idenokuchi fault was found east of Ono (around Loc. 245; Fig. 12.2), where a paved road was damaged with a small amount of vertical drop ($\sim$15 cm) down to the north (Fig. 12.5). From here to the northeast, no ground deformation indicative of surface faulting was found for a distance of $\sim$800 m, but the surface rupture reappeared immediately north of Shiraito Waterfall at Taki (Locs. 246 and 261). A 50–100-cm-high northwest-facing fresh fault scarp was continuously traced in the forest from Taki to Ioji (Locs 263–280; Fig. 12.6), along which the strike of the surface rupture slightly changed from NE to ENE. The largest measured vertical displacement along this stretch of the

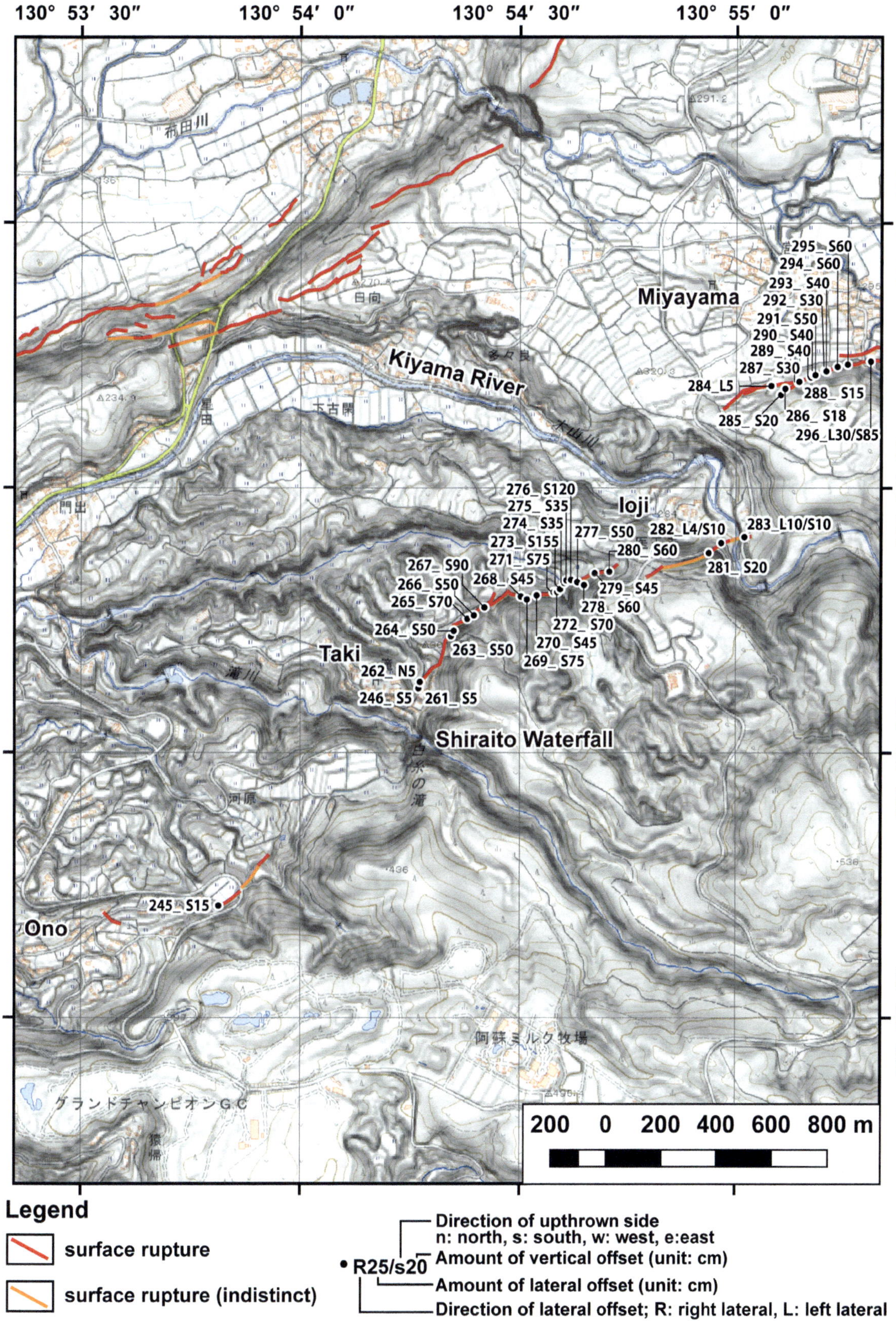

Fig. 12.2 Detailed surface rupture map of the southwestern part of the Idenokuchi fault section. Strike-slip and vertical displacements at each locality are also shown. The background is a half-transparent topographic map of Geospatial Information Authority of Japan (GSI) underlain by a slope map produced from 5 m grid digital elevation models (DEMs) of GSI

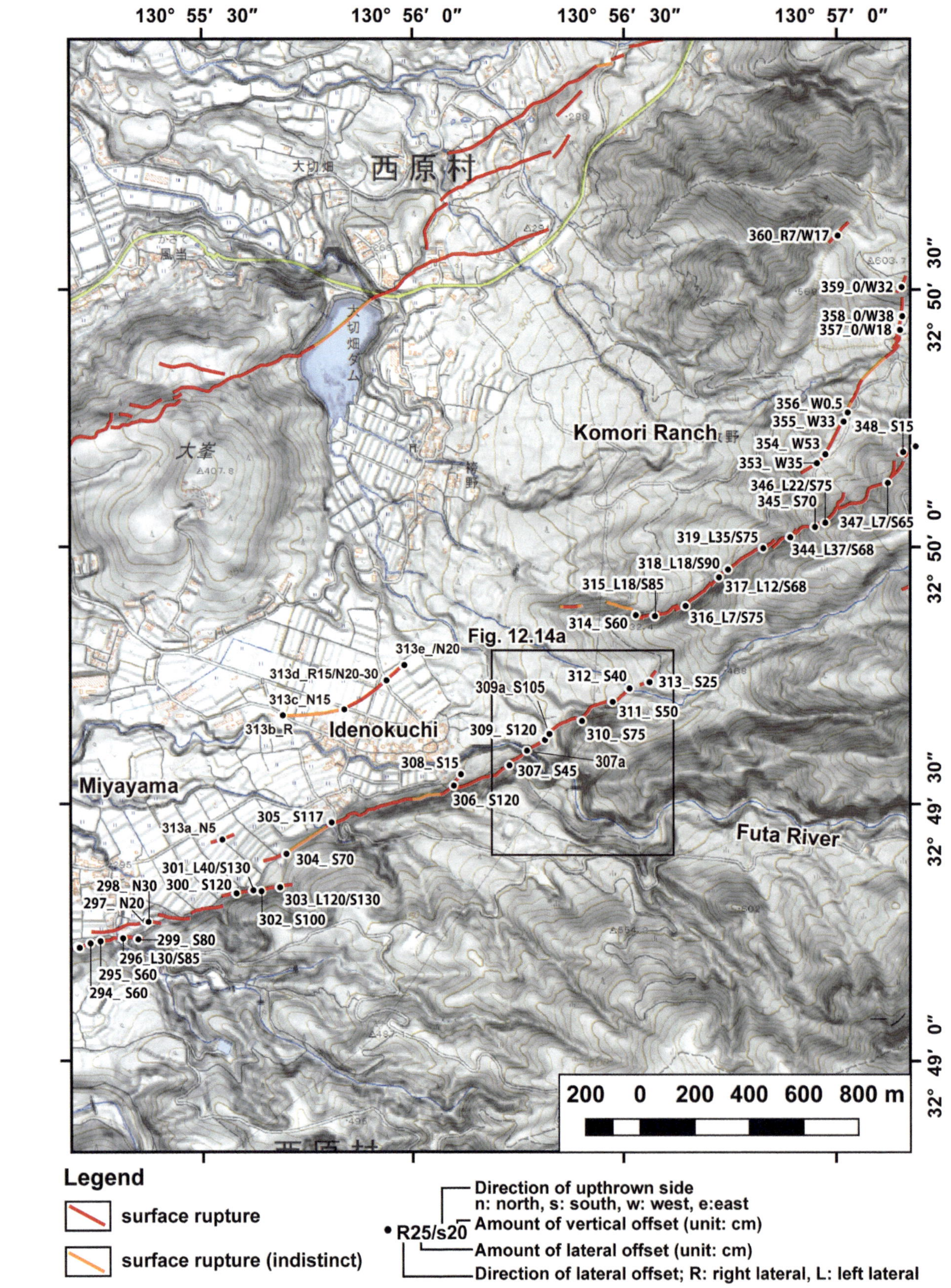

Fig. 12.3 Detailed surface rupture map of the middle part of the Idenokuchi fault section. Strike-slip and vertical displacements at each locality are also shown. The base map is the same as shown in Fig. 12.2

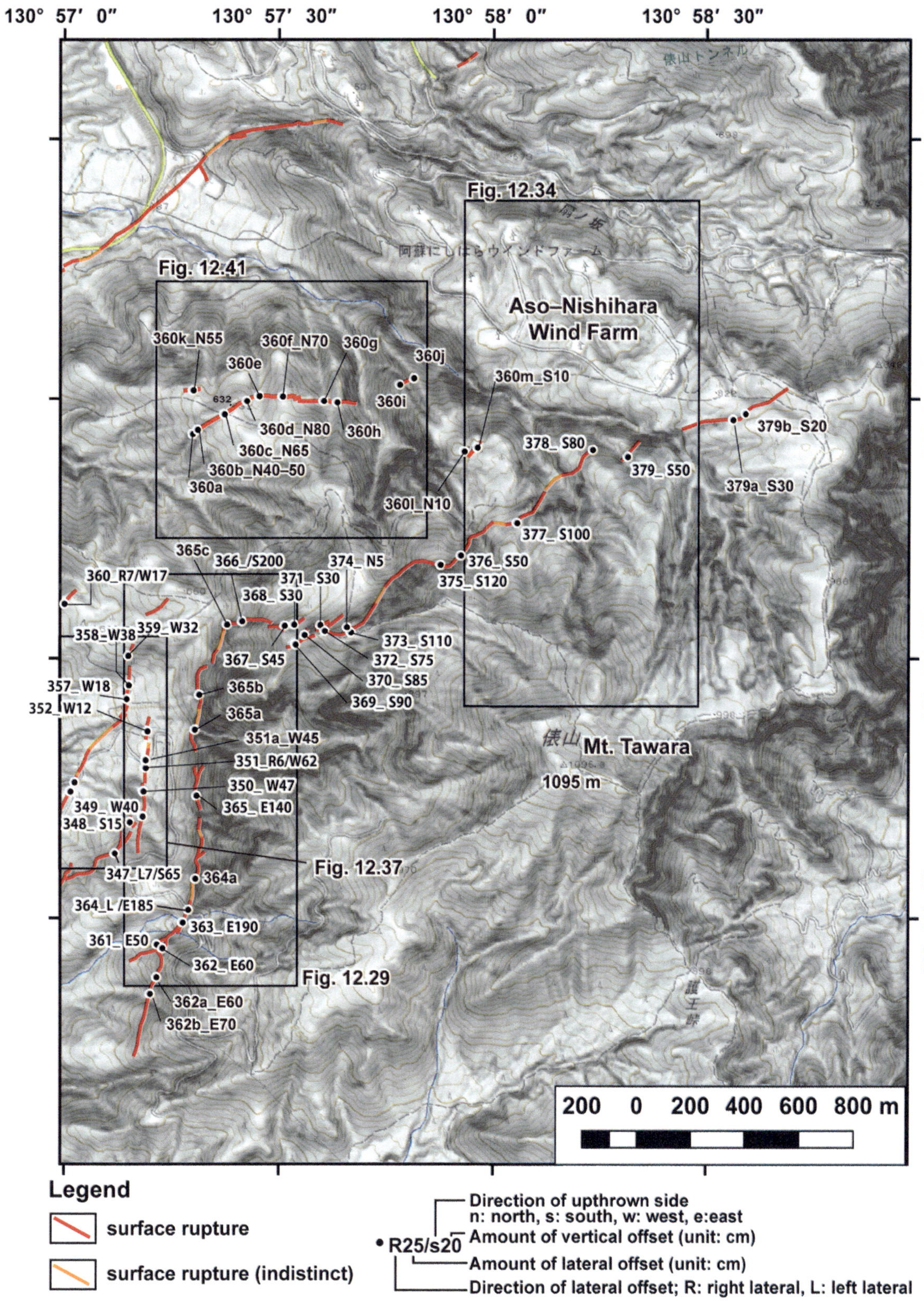

Fig. 12.4 Detailed surface rupture map of the northeastern part of the Idenokuchi fault section. Strike-slip and vertical displacements at each locality are also shown. The base map is the same as shown in Fig. 12.2

Fig. 12.5 Photograph of a damaged paved road west of Ono. Looking to the southwest. The location is ~100 m northeast of Loc. 245, where a vertical displacement was ~15 cm down to the north

Fig. 12.6 Photograph of a fresh fault scarp in the forest around Loc. 272. Looking to the northeast. The yellow-colored scale is 60 cm long

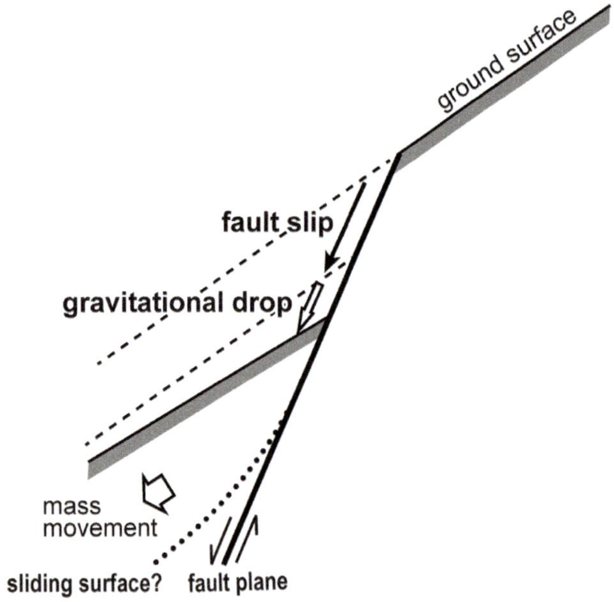

Fig. 12.7 Schematic illustration of local gravitational enhancement of vertical displacement

Fig. 12.8 Photograph of a damaged paved road at Loc. 282 south of Ioji. Looking to the southeast. The vertical displacement was measured at ~10 cm down to the north

fault was 155 cm (Loc. 273), but the displacement may have been locally enhanced due to gravitational mass movement that utilized the fault plane, as schematically illustrated in Fig. 12.7. The apparent gravitational enhancement of the vertical displacement was observed at many places along the Idenokuchi fault, notably where the fault cut through a steep mountain slope.

South of Ioji, the surface rupture made a small right stepover between Locs. 280 and 281. The vertical displacement becomes substantially smaller to the east of this stepover (10–20 cm; Fig. 12.8), and the fault scarp died out east of Loc. 283.

12.2.2 Tectonic Geomorphology

The surface ruptures from Ono to Ioji appeared along the base of a clear 50–100-m-high and northwest-facing escarpment that bounds hilly mountains composed of the middle Pleistocene pre-Aso volcanic rocks and 270–120 ka Aso-1–3 pyroclastic flow deposits to the southeast and well-dissected hills topped with Aso-4 pyroclastic flow deposits of ~90 ka to the northwest (Hoshizumi et al. 2004; Kumamoto Prefecture Geological Map Compilation Committee 2008). An inferred active fault trace had been mapped here on the preexisting active fault maps (Research Group for Active Tectonics in Kyushu 1989; Ikeda et al. 2001).

12.3 From Miyayama to Idenokuchi

12.3.1 Surface Ruptures

The surface rupture made a ~400-m-wide left stepover across the Kiyama River and reappeared along a middle to upper part of a preexisting fault scarp on the fluvial terrace surface south of Miyayama, which had been mapped before the earthquake (Research Group for Active Tectonics in Kyushu 1989; Ikeda et al. 2001). The WNW-facing co-seismic fault scarp increased its height eastward and crossed a curved paved road twice at Loc. 296, where a vertical displacement was measured at 85 cm (Figs. 12.3 and 12.9). The road was left-laterally offset by ~30 cm, but at least some of this strike-slip displacement would be an apparent effect because the fault crossed the road at a low angle here: horizontal extension due to normal faulting produces an apparent lateral offset on a linear feature that obliquely crosses the fault trace. Immediately to the north of this fault trace, a minor, 20–30-cm-high antithetic fault scarp appeared subparallel to the main fault scarp (Locs. 297 and 298), forming a local graben. The main fault scarp grew to the east, making small left stepovers, and reached 100–130 cm in height at Loc. 300–303. At Loc. 301, the ~130-cm-high fault scarp obliquely crossed and destroyed a buried water pipe (Fig. 12.10). A left-lateral displacement of ~40 cm was observed here, but again, at least some of this offset would be an apparent effect. At Loc. 303, a fault plane was exposed on a face of the ~130-cm-high fault scarp with a strike and dip of N85°E and 82°N, respectively. The slickenline was preserved on the fault plane and had a rake of 40–45° from the west. Assuming that the slickenline represents the overall slip vector at this site, the left-lateral displacement is estimated at 130–150 cm. Although the displacement appears too large and may indicate that the preserved slickenline only records the last-moment slip vector, this observation, along with left-lateral offset observed at Locs. 296 and 301, suggests that the slip around here was accompanied by some left-lateral component.

The surface rupture made a 100-m-wide left stepover between Locs. 303 and 304, and to the east, a fresh fault scarp could be traced nearly continuously along the base of a north-facing steep slope for a distance of ~1.6 km. At Loc. 305, a paved road and the concrete river channel immediately to the west were both destroyed and vertically displaced down to the north, with an amount of displacement being measured at ~120 cm on the road using a hand level

Fig. 12.9 Photograph of a ~85-cm-high fault scarp that crossed a small paved road twice at Loc. 296 south of Miyayama. Red arrows indicate the fault trace. The road was also left-laterally offset by ~30 cm, but at least some of this strike-slip displacement would be an apparent effect because the fault crossed the road at a low angle

Fig. 12.10 Photograph of a buried water pipe destroyed and displaced by the fault scarp at Loc. 301. The vertical displacement was ~130 cm. The scale in the middle of the photograph is 2 m long

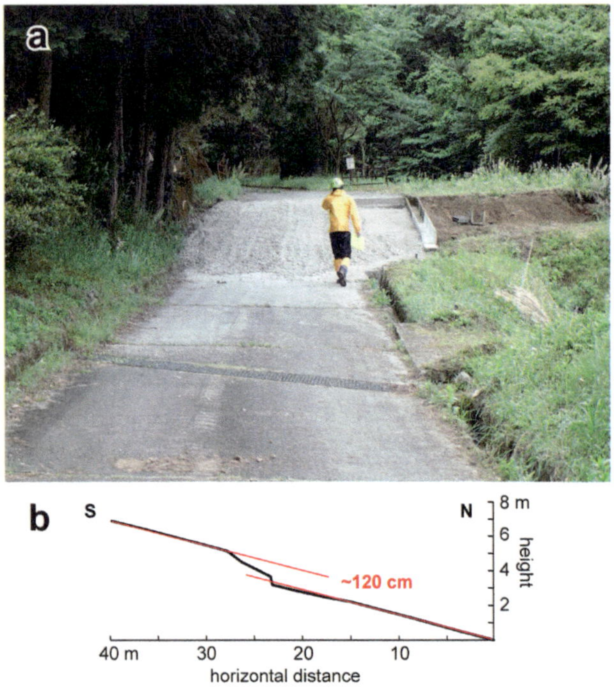

Fig. 12.11 **a** Photograph of a damaged paved road at Loc. 305 southwest of Idenokuchi. Looking to the south. **b** Topographic profile across the fault scarp at Loc. 305. A hand level and tape measure were used for the survey

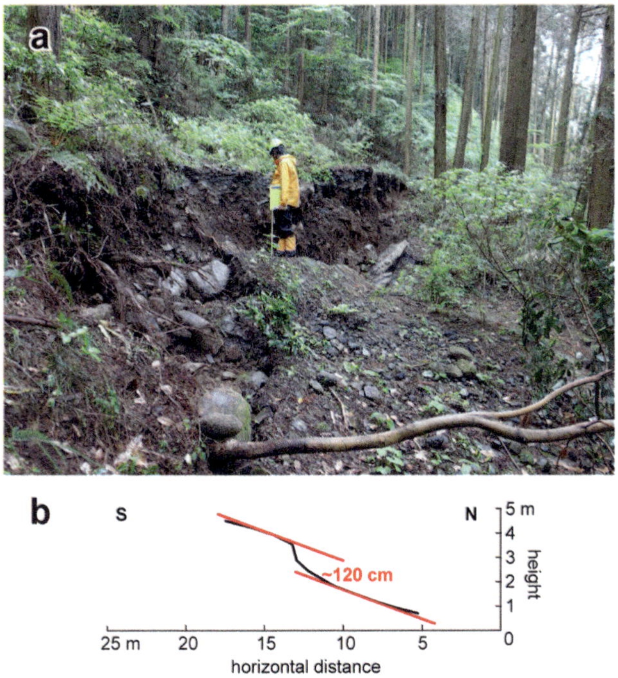

Fig. 12.12 **a** Photograph of a fault scarp that cut across an old, seemingly unused paved road at Loc. 306 south of Idenokuchi. Looking to the southwest. **b** Topographic profile across the fault scarp at Loc. 306. A high-accuracy laser distant meter (Laser Technology TruPulse 200X) was used for the survey

Fig. 12.13 Photograph of a fault scarp that crossed the bottom of the Futa River gorge at Loc. 307a. Looking to the southeast. The photograph was taken in March 2017, ~11 months after the earthquake. The scarp (indicated by red arrows) was composed of orange-colored fault breccia probably originated from pyroclastic deposits and was obviously modified and eroded by fluvial processes in the rainy season after the earthquake

and tape measure (Fig. 12.11). To the east, the fault scarp continued into the forest and crossed an old, seemingly unused paved road at Loc. 306 south of Idenokuchi (Fig. 12.12a). Topographic profiling with a high-accuracy laser distant meter (Laser Technology TruPulse 200X) suggested a vertical displacement of ~120 cm again here (Fig. 12.12b). A subsidiary fault trace immediately north of this site cut across the same road and formed a ~15-cm-high, north-facing fault scarp (Loc. 308).

The fault crossed a deep gorge of the Futa River (Futagawa) at Loc. 307a. We could not access the bottom of the gorge during the post-earthquake surface rupture mapping campaign because of extensive slope failures on both valley walls. In March 2017, ~11 months after the earthquake, however, we were able to reach the valley bottom and found that a zone of orange-colored fault breccia, which is probably originated from pyroclastic deposits, crossed the riverbed, and formed a clear northwest-facing fault scarp (Fig. 12.13). The scarp was several tens of centimeters high, but it was clear that the riverbed had been substantially modified and eroded by fluvial processes in the rainy season after the earthquake.

North of the Futa River, the fault scarp appeared at the base of a small isolated ridge to the southeast and crossed a dirt road at Loc. 309 (Figs. 12.14a and 12.15). On the road, the displacement was accommodated by ~95-cm-high main

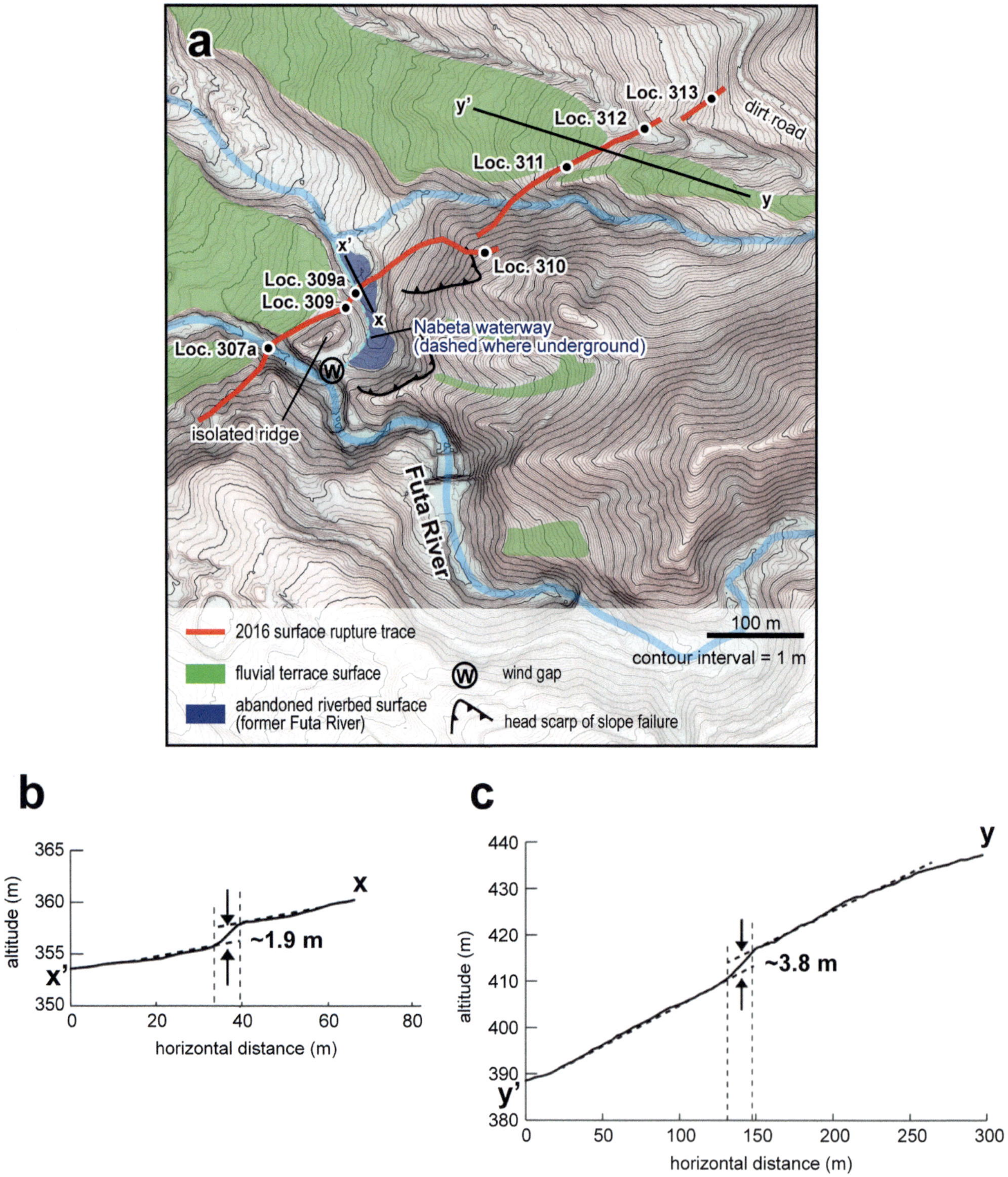

Fig. 12.14 **a** Detailed geomorphological map from Locs. 307a to 313. The background is a morphometric protection index red relief image map (MPI-RRIM; Kaneda and Chiba 2019) overlain by 1 m interval topographic contours, both of which are produced from post-earthquake 2 m grid DEMs by airborne light detection and ranging (LiDAR) surveys provided by GSI. See Fig. 12.3 for location. **b** Topographic profile along x–x′ line in the panel **a** across a preexisting fault scarp on the abandoned riverbed surface. The profile shows a vertical displacement of ~1.9 m, which is about twice as large as that in 2016. **c** Topographic profile along y–y′ line in the panel **a** across a preexisting fault scarp on a fluvial terrace surface. The profile shows a vertical displacement of ~3.8 m, which is 7–8 times larger than that in 2016. The topographic profiles in panels **b** and **c** are produced with post-earthquake 2 m grid LiDAR DEMs

Fig. 12.15 Photograph of a fault scarp that crossed a dirt road at Loc. 309. Looking to the southeast. The displacement was accommodated by a ~95-cm-high main and ~25-cm-high subsidiary scarps, amounting to a total vertical displacement of ~120 cm

and ~25-cm-high subsidiary scarps, amounting to a total vertical displacement of ~120 cm here. The main scarp exposed a beautiful planar surface, probably a fault plane, which had a strike and dip of N52° E and 64° N, respectively.

To the northeast, the fault scarp extended down onto an abandoned riverbed at Loc. 309a. The Futa River at present flows through the aforementioned gorge at Loc. 307a, but is interpreted to have taken another course until sometime and flowed on this abandoned surface via a wind gap (circled W in Fig. 12.14a). Now, an artificial Nabeta waterway goes along the western margin of the abandoned surface, the southern half of which is an underground channel (solid and dashed light blue lines in Fig. 12.14a). The 2016 rupture appeared in this waterway tunnel at ~5 m south of the tunnel entrance, destroying the waterway bed with a vertical displacement of ~105 cm (Fig. 12.16a, b). A fresh fault scarp was also found on a mountain slope above the underground waterway (Fig. 12.16a), and the locations of those underground and surface ruptures allowed a rough estimate of a near-surface fault dip (60° N; Toda et al. 2016; Fig. 12.16b). About 4 years later, in 2020, this part of the slope was entirely excavated for channel restoration, and a clear fault was exposed on the sidewall of the excavation (Fig. 12.16c). The fault plane had a strike of N58–63°E and a dip of 66° N and separated tuff breccia on the upthrown (southern) side and blackish and yellowish-brown surface soils on the downthrown (northern) side.

An interesting observation here is the presence of a preexisting fault scarp on this abandoned riverbed, and the scarp grew during the 2016 earthquake (Fig. 12.17). Topographic profile across the fault scarp from post-earthquake 2 m grid digital elevation models (DEMs) by airborne light detection and ranging (LiDAR) surveys shows a vertical displacement of ~1.9 m (Fig. 12.14b), which is substantially larger than that by the 2016 earthquake at Loc. 309a (105 cm). The cumulative displacement might be even larger given that the downthrown side is more or less buried by tributary deposits from the east. Our pilot hand cores on the upthrown side reveal that the surface is overlain only by 50–60-cm-thick black and brown soil that does not include Kikai-Akahoya (K-Ah) tephra of ~7.3 ka based on a 5 cm interval crypto-tephra analysis. This observation suggests that the penultimate surface faulting event on this fault postdates ~7.3 ka.

The fault scarp continued to the northeast, hiked up a mountain slope, and made a small left stepover after Loc. 310. The scarp then crossed a small dry stream and further extended onto a fluvial terrace surface, on which a trail was vertically displaced down to the northwest by ~50 cm at Loc. 311 (Fig. 12.14a). On this terrace surface, a preexisting fault scarp was visible on aerial photographs taken in the 1960s and 1970s, but this feature had not been mapped on preexisting active fault maps (Research Group for Active Tectonics in Kyushu 1989). Although the age of the surface is unknown, a topographic profile across the fault scarp using post-earthquake 2 m grid LiDAR DEMs suggests a vertical displacement of ~3.8 m (Fig. 12.14c), which is 7–8 times larger than the height of the 2016 scarp, indicating recurrent surface faulting along this fault trace.

At a dry stream north of this terrace surface, the surface rupture ran through a small dam built of stones (Loc. 312). The dam appeared to have been forcibly uplifted, as the downstream side face of the dam structure was significantly cracked at its base (Fig. 12.18). From the width of the open crack, the amount of dam uplift or the vertical fault displacement was estimated to be at least ~40 cm. After damaging the dam, the fault made a small right stepover and died out around northeast of Loc. 313. No clear evidence for surface faulting was found along a dirt road in the Komori Ranch further to the northeast (Figs. 12.3 and 12.14a).

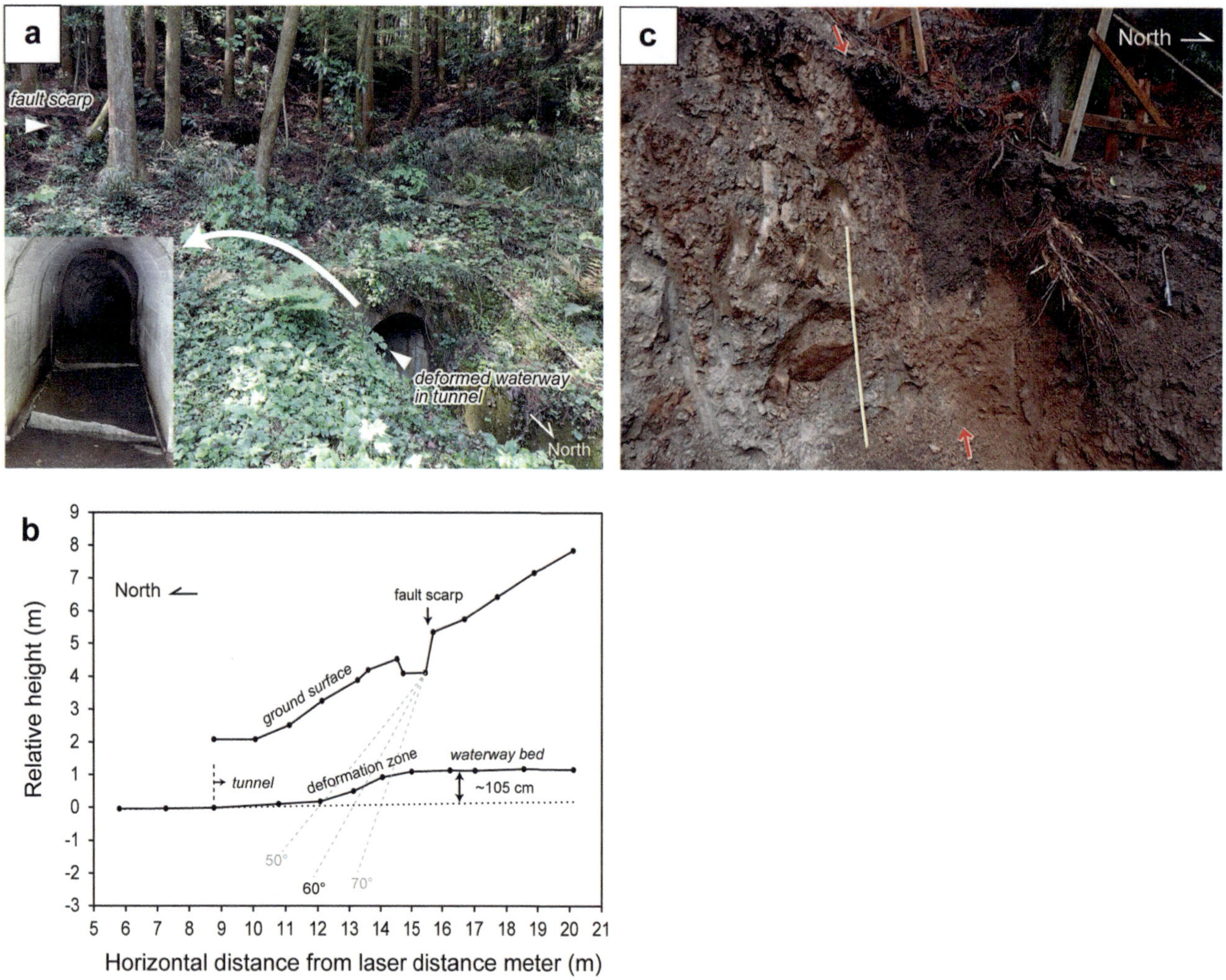

Fig. 12.16 **a** Photograph of a fault scarp on a mountain slope, beneath which the waterway tunnel was deformed at Loc. 309a. Looking to the southeast. Inset shows a photograph of the deformed waterway bed in the tunnel. **b** Profiles of the ground surface and waterway bed based on a laser distance meter survey. Dashed gray lines indicate possible surface fault dips, and a dip of 60° best fits the locations of the fault scarp and the underground deformation zone. **c** Photograph of a clear fault plane that separates tuff breccia on the upthrown side (left) and blackish and yellowish-brown surface soils on the downthrown side (right). Looking to the southwest. The yellow scale to the left of the fault is 1 m long. Red arrows indicate the fault plane. The fault was exposed when the fault scarp on the mountain slope was entirely excavated for channel restoration in 2020. The panels **a** and **b** are modified after Fig. 5 of Toda et al. (2016)

Fig. 12.17 Photograph of a fault scarp that cut across an abandoned former riverbed of the Futa River east of Loc. 309a. Looking to the east. Red arrows indicate the fault trace. It appears that a preexisting fault scarp grew during the 2016 earthquake

Fig. 12.18 Photograph of a cracked and uplifted stone dam at Loc. 312. Looking to the east. Red arrows indicate the fault trace

Fig. 12.19 Photograph of a stratigraphic section beneath a minor antithetic surface rupture north of Idenokuchi at Loc. 313d. Looking to the northeast. A yellowish-brown layer of K-Ah tephra (~7.3 ka; indicated by the yellow arrow) is folded up and becomes nearly vertical beneath where the road pavement was cracked due to a 20–30-cm-high monoclinal deformation associated with the 2016 earthquake. The red arrow shows the location of the road cracks. Vertical and horizontal scales on the left of the photograph are both 1 m long. The section was exposed associated with road reconstruction practices in 2019

From Miyayama to Idenokuchi, two minor antithetic fault traces were also found ~200–500 m north of the main fault scarp (Fig. 12.3). East of Miyayama, ~70-m-long surface rupture cut through rice paddies with a small (~5 cm) amount of vertical displacement down to the south (Loc. 313a). Similarly, ~500-m-long minor surface rupture appeared across roads and rice paddies north of Idenokuchi (Locs. 313b–313e). The vertical displacement was 15–30 cm down to the south, and this fault trace was associated with a small right-lateral slip less than 15 cm. During road reconstruction practices in 2019, excavation was conducted across the fault trace down to a depth of ~2 m at Loc. 313d. The exposed stratigraphic section showed that a yellowish-brown K-Ah tephra layer (~7.3 ka) was folded up and became nearly vertical beneath where the road pavement was cracked due to a 20–30-cm-high monoclinal deformation in 2016 (Fig. 12.19). This observation indicates repetition of faulting even along this minor antithetic fault trace.

12.3.2 Tectonic Geomorphology

The fault trace from Miyayama to Idenokuchi was characterized by a clear topographic boundary between hilly terrains composed of middle Pleistocene pre-Aso volcanic rocks to the southeast and Late Pleistocene volcanic fan or talus surfaces to the northwest (Hoshizumi et al. 2004; Kumamoto Prefecture Geological Map Compilation Committee 2008). In particular, there are preexisting synthetic and antithetic fault scarps that cut across a fan surface at Miyayama, and based on this tectonic-geomorphic feature, the active Idenokuchi fault had been mapped on published active fault maps along this section of the fault (Research Group for Active Tectonics in Kyushu 1989; Ikeda et al. 2001). However, the mapped active fault is up to the Futa River, and preexisting fault scarps on fluvial terrace surfaces to the northeast, shown in Fig. 12.14, had not been identified before the earthquake.

12.4 Main Fault Scarp in the Komori Ranch

12.4.1 Surface Ruptures

A clear normal fault scarp continuously appeared across grassy mountain slopes of the Komori Ranch for a distance of ~1.4 km (Locs. 314–319 and 344–348; Fig. 12.3), which was visible even from a distant piedmont area (Fig. 12.20). The scarp may appear to be a head scarp of a large-scale landslide, but the scarp height was quite stable in the range of 60–90 cm for a long distance across ridges and valleys, suggesting that the scarp was tectonic in origin. Gravitational scarps are generally shorter and associated with a much higher variation in their heights. The tectonic origin of the scarp is also supported by the fact that it is located on the northeastern extension of the typical normal fault surface ruptures from Miyayama to Idenokuchi as described in the last section, and that there is consistent left-lateral slip component of less than 40 cm throughout its trace.

The fault scarp crossed a small paved road in the Komori Ranch at Loc. 314 (Fig. 12.21), where topographic profiling using a laser distance meter suggested a vertical displacement of ~60 cm. The scarp could be traced westward for a short distance, but soon became unclear. To the east, the surface rupture obliquely hiked up a grassy mountain slope. The fault trace made a smooth curve to change its strike from east to northeast around Locs. 315–316. At many places along this fault stretch, jigsaw matching of irregularly torn grass edges on the upthrown and downthrown sides allowed a rough estimate of the amount of strike-slip offset in addition to the vertical displacement. At Loc. 315, the left-lateral offset of 18 cm was measured in addition to the vertical displacement of 85 cm. Similarly, the 7 cm left-lateral offset was associated with the 75 cm vertical displacement at Loc. 316. The consistent left-lateral slip component along this

Fig. 12.20 Distant-view photograph of a normal fault scarp in the Komori Ranch. Looking to the east from the north of Idenokuchi. Red arrows indicate the fault trace

Fig. 12.21 Photograph of a fault scarp that crossed a small paved road in the Komori Ranch at Loc. 314. Looking to the southwest. The vertical displacement was measured at ~60 cm

Fig. 12.22 Photograph of a fault scarp that cut across a grassy mountain slope of the Komori Ranch at Loc. 318. Looking to the southwest. The scarp height was 90 cm, the largest vertical displacement measured along this fault strand

fault stretch was also inferred from frequent small right stepovers in the fault trace.

The fault scarp further hiked up the mountain slope up to a ridge at Loc. 319. On the way to the ridge, the scarp height was 90 cm at Loc. 318 (Fig. 12.22), the largest vertical displacement along this fault strand. From Loc. 319, the surface rupture went down to a valley bottom and again hiked up a mountain slope up to around Loc. 346, where the vertical and left-lateral displacements were 75 cm and 22 cm, respectively. Further to the northeast, the surface rupture extended across three small valleys and reached Loc. 347, the highest point along this fault trace at an altitude of 610 m. Northeast of Loc. 347, the surface rupture branched into two traces, and both fault scarps soon became unclear and died out. On the northern extension, we found a 15-cm-high fault scarp at Loc. 348, but this scarp did not extend to the northeast or southwest.

12.4.2 Tectonic Geomorphology

This stretch of the surface rupture in the Komori Ranch appeared in the middle to lower-middle part of the preexisting ~10–40 m high much larger scarp visible on aerial photographs taken in the 1960s and 1970s. The cumulative scarp displaces mountain slopes composed of middle Pleistocene pre-Aso volcanic rocks (Hoshizumi et al. 2004; Kumamoto Prefecture Geological Map Compilation Committee 2008), but no clear geomorphic evidence for late Pleistocene to Holocene activity can be identified along this fault strand. In addition, the scarp might be mistaken as a crown scarp of a large-scale landslide because the map-view trace is specifically curved in its south-westernmost part (Fig. 12.3). Accordingly, no active fault had been mapped before the earthquake along this section of the fault (Research Group for Active Tectonics in Kyushu 1989; Ikeda et al. 2001).

12.5 Western Foot of Mt. Tawara

12.5.1 Surface Ruptures

To the east of the main fault scarp in the Komori Ranch mentioned earlier, we found another even more extensive normal fault scarp that extended for a distance of ~1.6 km along the western foot of the 1095-m-high Mt. Tawara (Tawara-yama) (Fig. 12.4; Locs. 362b–365c).

At Loc. 364, the surface rupture cut across a dirt road, forming a spectacular west-facing normal fault scarp that well exceeded human height (Fig. 12.23a). Topographic profiling with a laser distance meter suggested a vertical displacement of 185 cm here (Fig. 12.23b). The displacement may look a bit small compared to the appearance of the photograph. This is because small antithetic scarps occurred immediately to the west of the main scarp, forming a small local graben at the scarp base (blue plots in Fig. 12.23b), thereby exaggerating the fault displacement. A part of the scarp face exhibited a quite smooth planar surface, most likely a fault plane, which had a strike of N18° E and a dip of 72° W (Fig. 12.23c). Clear slickenlines were also visible on the plane with a rake being 83° from the south, suggesting a small left-lateral slip component here. The impressive fault scarp at Loc. 364 was later affected by a local slope failure. When we revisited here 10 months later, in March 2017, the downthrown side of the scarp had been destroyed and covered by landslide debris, although the scarp was still clearly visible (Fig. 12.23d).

From Loc. 364 to the south, the fault scarp went down to the bottom of a small valley and formed a scarp of similar height (~190 cm) (Loc. 363). The surface rupture could be traced to the south, crossed another small valley, and then branched into two subparallel traces. Both fault scarps traversed the same shallow gully at Locs. 361 and 362

Fig. 12.23 **a** Photograph of a spectacular normal fault scarp that cut across a dirt road at Loc. 364, western foot of Mt. Tawara. Looking to the east. This is the original unclipped photograph of Fig. 4 of Toda et al. (2016). **b** Topographic profile across the fault scarp at Loc. 364. A high-accuracy laser distant meter (Laser Technology TruPulse 200X) was used for the survey. **c** Photograph of a smooth planar surface, most likely a fault plane, exposed on the face of the fault scarp at Loc. 364. The pen on the plane pointed in the direction of slickenlines. Inset shows a close-up view of the slickenlines. **d** Photograph of the same fault scarp of the panel **a** but taken 10 months later. The downthrown side of the scarp was destroyed and covered by debris from a local slope failure, although the scarp was still clearly visible

(Fig. 12.24), the height of which was measured at ~50 cm and ~60 cm, respectively. The two surface ruptures soon remerged into a single fault trace, but the fault scarp became unclear and died out ~100 m southwest of Locs. 361 and 362.

There was another surface rupture trace that directly hiked up the mountain slope from southwest of Locs. 361 and 362. This trace continued to the south for a distance of ~500 m, making multiple large and small right stepovers and crossing several small gullies, and died out north of a deep valley. The height of the fault scarp was ~60 cm and ~70 cm at Locs. 362a and 362b, respectively, and gradually decreased toward the south.

Back to Loc. 364 on the dirt road, the fault scarp went up the mountain slope and continued to the north for a distance of ~1 km, principally along a north–south trending clear topographic boundary between relatively gentle alluvial-cone or talus surfaces to the west and steep mountain slopes of Mt. Tawara to the east. One important feature of this stretch of the fault is that gravitational movement appeared to be overprinted on tectonic displacement at many places, possibly resulting in substantial displacement

Fig. 12.24 Photographs of two subparallel fault scarps that cross the same shallow gully at Loc. 361 (**a**) and Loc. 362 (**b**). The two sites were ~25 m apart with Loc. 361 located downstream. The scarp height was ~50 cm and ~60 cm at Locs. 361 and 362, respectively. Yellow arrows show the direction of the gully characterized by the gravelly ground surface. Red arrows indicate the fault trace

Fig. 12.25 Photograph of a fresh fault scarp that cut across a steep mountain slope and extended subparallel to topographic contours at Loc. 364a. Looking to the north

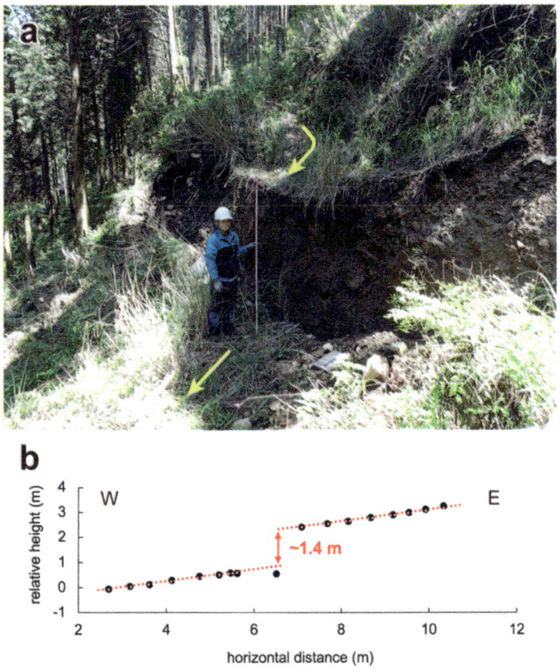

Fig. 12.26 **a** Photograph of a nearly vertical fault scarp that cut across a small abandoned trail at Loc. 365. Looking to the north. Yellow arrows indicate the abandoned trail. **b** Topographic profile across the fault scarp at Loc. 365. A high-accuracy laser distant meter (Laser Technology TruPulse 200X) was used for the survey

exaggeration as schematically illustrated in Fig. 12.7, the gravitational retreat of a fault scarp, and the mixture of the both effects. The complex tectonic and gravitational processes made it difficult to confidently measure a fault displacement along this stretch of the fault despite its striking appearances. Also, many large and small slope failures completely destroyed or buried the surface ruptures along this fault stretch.

At Loc. 364a, ~100 m north of Loc. 364, a fresh fault scarp cut across a steep mountain slope and extended subparallel to topographic contours, where gravitational effect seemed to be minimal (Fig. 12.25). To the north, the surface rupture crossed a small abandoned trail at Loc. 365 and formed a nearly vertical fault scarp (Fig. 12.26a). Topographic profiling with a laser distance meter suggested a vertical displacement of 140 cm here (Fig. 12.26b).

Further to the north, a pair of large-scale slope failures crossed the surface rupture and exposed spectacular fault planes between Locs. 365a and 365b. In particular, a huge fault scarp traversed the northern slope failure and seemed like a big wall viewed from far below (Fig. 12.27a). The height of the scarp was as high as 10–11 m based on rough remote leveling using a laser distance meter (Fig. 12.7b, c), suggesting that the fault scarp was obviously and substantially exaggerated by gravitational movement on the downthrown side as schematically illustrated in Fig. 12.7. The scarp face was quite smooth and planar and thus interpreted as a fault plane. The strike and dip of the plane were N22° W and 60°W, respectively, but the slickenlines were not clear.

Though partially disrupted by small slope failures, the fault scarp further continued to the north and reached Loc.

Fig. 12.27 **a** Distant-view photograph of a huge fault scarp that traversed across a large slope failure south of Loc. 365b. Looking to the east. Red arrows indicate the fault scarp. **b** Photograph of the same fault scarp in the panel **a** taken from immediate south. Looking to the north. **c** Photograph of the same fault scarp in the panel **a** taken from immediate north. Looking to the south. The height of the scarp was as high as 10–11 m based on rough remote leveling using a laser distance meter, suggesting that the fault scarp was obviously and substantially exaggerated by gravitational movement on the downthrown side

Fig. 12.28 Photograph of a fault scarp at Loc. 365c, where the surface rupture made an abrupt turn to leave the western foot of Mt. Tawara and began to hike up the mountain to the east. Looking to the southeast

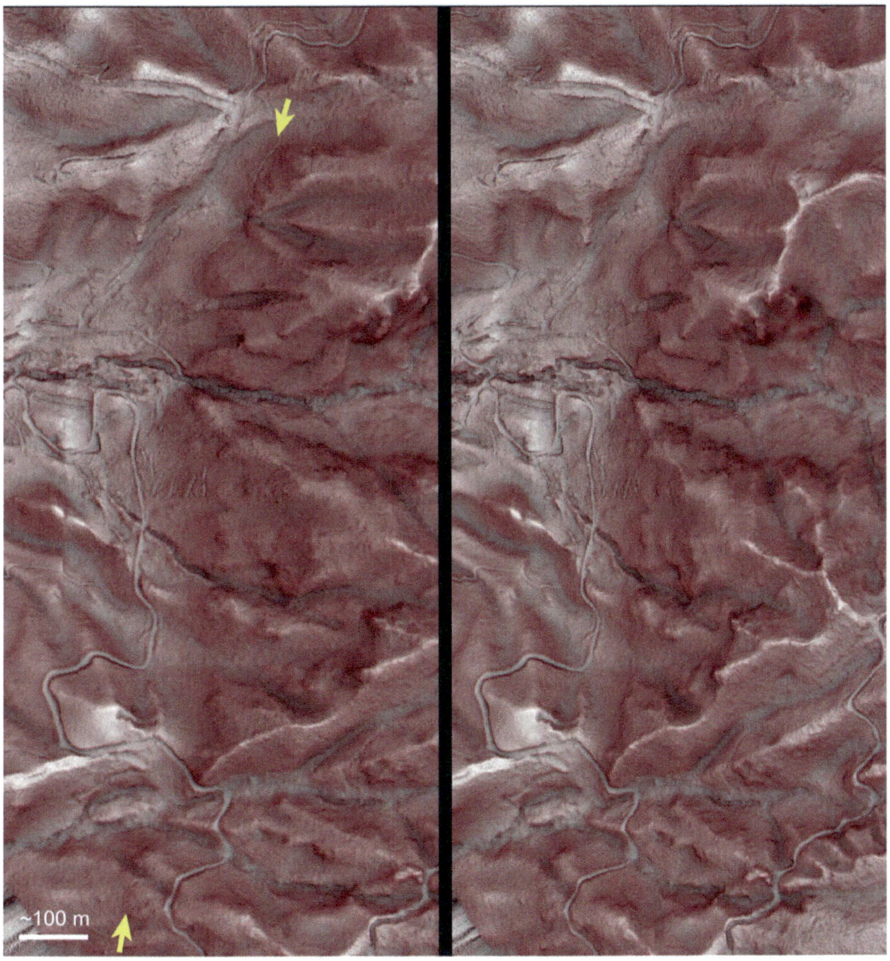

Fig. 12.29 Stereopaired morphometric protection index red relief image maps (Stereo MPI-RRIMs; Kaneda and Chiba 2019) made from the pre-earthquake 1 m grid LiDAR DEMs along the western foot of Mt. Tawara. Note that relatively fresh scarps (indicated by yellow arrows) intermittently extend along a clear topographic boundary between alluvial-cone or talus surfaces to the west (left) and steep mountain slopes to the east (right). The DEMs were provided by GSI. See Fig. 12.4 for location

365c, where the surface rupture made an abrupt turn to leave the western foot of Mt. Tawara and began to hike up the mountain to the east. Here a fault scarp higher than human height exposed a smooth fault plane with a strike and dip being N42°E and 70°W, respectively (Fig. 12.28).

12.5.2 Tectonic Geomorphology

Although the maximum vertical displacement of the 2016 earthquake was recorded on this stretch of the fault, no active fault had been identified in the preexisting active fault map (Research Group for Active Tectonics in Kyushu 1989). However, the western foot of Mt. Tawara is characterized by a clear and linear topographic boundary between alluvial-cone or talus surfaces to the west and steep mountain slopes composed of middle Pleistocene pre-Aso volcanic rocks (Hoshizumi et al. 2004; Kumamoto Prefecture Geological Map Compilation Committee 2008) to the east. Looking closely at the image from the pre-earthquake 1 m grid LiDAR DEMs, one can further notice that the steep mountain slope is buttressed by a narrow zone of even steeper slopes or scarps (Fig. 12.29). The freshness and continuity of those scarps imply repetitive surface faulting along this topographic boundary in the recent past, probably in the Holocene time, although there is no direct information on the age of those geomorphic features.

12.6 Northwest of Mt. Tawara to Aso–Nishihara Wind Farm

12.6.1 Surface Ruptures

After the abrupt strike change at Loc. 365c, the 2016 surface rupture extended to the northeast for another ~2.4 km from the northwest of Mt. Tawara to the southeast of a windmill farm called Aso–Nishihara Wind Farm (Fig. 12.4).

To the west of Loc. 365c, the fault scarp extended to the east and directly hiked up the mountain slope. The height of the scarp is still as large as ~200 cm at Loc. 366 (Fig. 12.30), but it may have been exaggerated by gravitational movement on the downthrown side as the scarp was on a north-facing steep slope. The scarp continued up and made a small right stepover, but its height suddenly decreased toward Locs. 367 and 368, where the vertical displacement was 45 cm and 30 cm, respectively. The scarp soon died out.

A clear fault scarp reappeared across a relatively flat or gentle surface on a ridge. The height of the scarp was quite stable along this fault trace, implying minimal gravitational disturbances, and was measured to be 90 cm at Loc. 369 and 85 cm at Loc. 370. The tilting of trees was evident on the upthrown side (Fig. 12.31).

From Loc. 372 to the northeast, a northwest-facing fault scarp could be traced nearly continuously for a distance of ~1.2 km across ridges and valleys up to Loc. 378 (Fig. 12.4). The map-view trace was characterized by quite systematic bends; convex to the southeast on a ridge and convex to the northwest in a valley, suggesting normal faulting on a fault plane that steeply dips to the northwest. The maximum scarp height along this section of the fault was 120 cm at Loc. 375 (Fig. 12.32), but the displacement was exaggerated by downslope gravitational movement at some places.

The fault scarp suddenly became unclear immediately northeast of Loc. 378, but it reappeared southwest of Loc. 379, forming a ~100-m-wide right stepover, where a vertical displacement was measured at ~50 cm (Fig. 12.4). The fault scarp again became unclear where the fault would extend across a deep valley. To the northeast, we found that a small northwest-facing fault scarp extended southeast of Aso–Nishihara Wind Farm. The vertical displacement was 30 cm at Loc. 379a and 20 cm at Loc. 379b (Fig. 12.33). The displacement quickly decreased to the northeast, and only intermittent open cracks were visible along the north-easternmost 100-m-long section. We were unable to find evidence for the surface rupture further to the northeast.

Fig. 12.30 Photograph of a fault scarp at Loc. 366. The person holds a 3-m-long staff. The height of the scarp was still as large as ~200 cm, but it may have been exaggerated by gravitational movement on the downthrown side as the scarp was on the north-facing steep slope. Looking to the west

Fig. 12.31 Photograph of a ~90-cm-high fault scarp that appeared on a relatively flat or gentle surface on a ridge at Loc. 369. The tilting of trees was evident on the upthrown side. The person holds a 2-m-long staff. Looking to the northeast

Fig. 12.32 Photograph of a ~120-cm-high fault scarp that cut across the mountain slope at Loc. 375 northwest of Mt. Tawara. The person holds a 3-m-long staff. Looking to the east-southeast

Fig. 12.33 Photograph of a ~20-cm-high small fault scarp that extended in the forest at Loc. 379b. The site was located on a northwest-sloping dissected surface, on which giant windmills of Aso–Nishihara Wind Farm scatter. Looking to the southwest

12.6.2 Tectonic Geomorphology

The surface ruptures of this section cut across the middle part of the northwestern slope of Mt. Tawara, and unlike the western foot of the mountain, no clear topographic boundary can be identified along this section (Fig. 12.4). Accordingly, no active fault had been mapped before the 2016 earthquake (Research Group for Active Tectonics in Kyushu 1989). Even with the pre-earthquake 1 m grid LiDAR DEMs, it is still difficult to identify clear geomorphic evidence for the

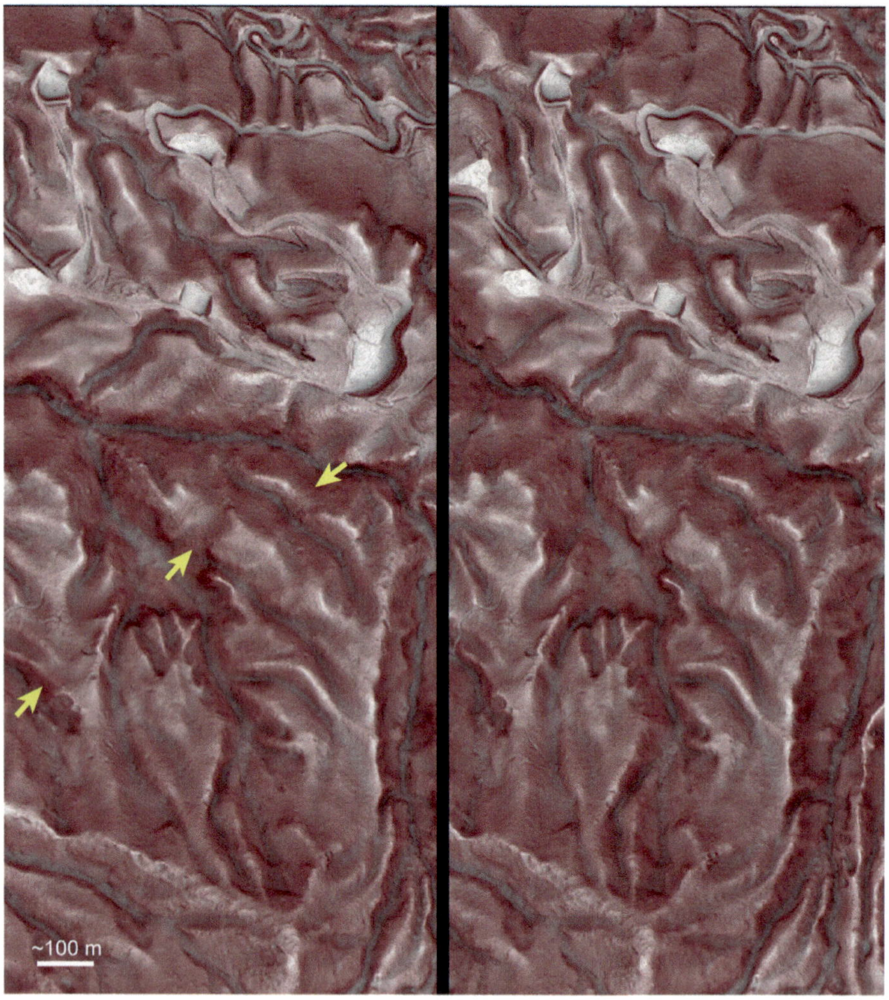

Fig. 12.34 Stereopaired morphometric protection index red relief image maps (Stereo MPI-RRIMs; Kaneda and Chiba 2019) made from the pre-earthquake 1 m grid LiDAR DEMs north of Mt. Tawara. Note the presence of preexisting fault scarps (indicated by yellow arrows) that appear to cut across a northwest-sloping well-dissected and degraded surface composed of 270–120 ka Aso-1–3 pyroclastic flow deposits (Hoshizumi et al. 2004). The DEMs were provided by GSI. See Fig. 12.4 for location

late Pleistocene or Holocene surface faulting along this section of the fault (Fig. 12.34). In the northeastern part, however, a northwest-sloping well-dissected and degraded surface reportedly composed of 270–120 ka Aso-1–3 pyroclastic flow deposits (Kumamoto Prefecture Geological Map Compilation Committee 2008) extends across the fault trace, and this surface appears to be vertically displaced by ~10 m along a limited section around Locs. 378 and 379 (Figs. 12.4 and 12.34).

12.7 Antithetic Fault Scarps in the Komori Ranch

12.7.1 Surface Ruptures

One interesting feature of the northern part of the Idenokuchi fault section is that it accompanied extensive antithetic (uphill-facing) surface faulting a few to several hundred meters west of the main surface rupture traces.

The most prominent antithetic faulting was found ~150–200 m northwest of and subparallel to the main normal fault scarp in the Komori Ranch (Figs. 12.3 and 12.4). An uphill-facing scarp nearly continuously extended from southwest of Loc. 353 up to the north of Loc. 359 for a distance of ~800 m with the maximum vertical displacement of 53 cm at Loc. 354. From Loc. 353 to the north of Loc. 356, the surface rupture appeared along the base of a prominent preexisting uphill-facing scarp on which a power transmission tower had been built (Fig. 12.35).

The surface rupture crossed and destroyed a paved road twice at Loc. 356 and immediately north of Loc. 356 and then became unclear to the northeast. The surface rupture reappeared in the forest south of Loc. 357, again crossing a paved road twice around here. To the north, an ~20–40-cm-high uphill-facing scarp extended across an expansive grass field on a fluvial terrace surface that slopes to the west (Locs. 357–358). It appeared that the terrace surface had been deformed before the earthquake such that the downstream side was uplifted relative to the upstream side,

Fig. 12.35 Photograph of a ~35-cm-high uphill-facing fault scarp at Loc. 353 that appeared along the base of a prominent preexisting larger scarp on which a power transmission tower had been built. Taken after Fig. 4e of Toda et al. (2016). Looking to the north

Fig. 12.36 Photograph of an ~20–40-cm-high uphill-facing scarp around Locs. 357–358 that extended across an expansive grass field on a fluvial terrace surface. Looking to the south. Note that the surface appears to have been deformed before the earthquake such that the downstream (right-hand) side was uplifted relative to the upstream (left-hand) side, forming a relatively flatter or dammed-up zone upstream (left) of the fault trace indicated by red arrows

forming a relatively flatter or dammed-up zone (Fig. 12.36). The pre-earthquake 1 m grid LiDAR DEMs confirm the preexisting deformation of the terrace surface (Fig. 12.37), based on which the cumulative vertical displacement of the surface was ~5–7 m before the 2016 earthquake. After deforming a paved road north of Loc. 359, the surface rupture made a slight left step, went up a steep slope, and soon became unclear before it reached a ridge to the north.

About 80–300 m east of the prominent uphill-facing fault scarp mentioned earlier, we found another trace of antithetic surface faulting for a distance of ~500 m from the south of Loc. 349 to the north of Loc. 352 (Fig. 12.4). Although not mapped before the earthquake, this surface rupture also occurred along a preexisting uphill-facing scarp (Fig. 12.37). In particular, a 47-cm-high uphill-facing scarp faithfully appeared at the base of a small knoll on a ridge at Loc. 350 (Fig. 12.38). The surface rupture continued to the north, with the scarp height being highest (62 cm) at Loc. 351, immediately north of which the surface rupture broke a paved road and formed a 45-cm-high steep ramp at Loc. 351a (Fig. 12.39). The scarp extended for another ~100 m to the north but became unclear in the forest north of Loc. 352. This antithetic surface rupture trace was located behind (east of) the main fault scarp found earlier in the Komori Ranch, having led us to assume the presence of another main surface rupture further to the east and to initiate a search along the western foot of Mt. Tawara, although no active fault had been mapped before the earthquake.

Another minor antithetic faulting was found at the northeastern margin of the Komori Ranch at Loc. 360 (Fig. 12.3). Here, an ~17-cm-high southeast-facing scarp cut across a small trail (Fig. 12.40) but only extended for a distance of ~100 m in the northeast–southwest direction. The surface rupture appeared along a preexisting, much higher uphill-facing scarp on the mountain slopes that extended for a much longer distance of at least 500–600 m. However, the preexisting scarp is more degraded than other preexisting antithetic scarps mentioned earlier, and only a part of it appears to have grown in 2016 as we were unable to find evidence of surface faulting along the rest of this preexisting scarp.

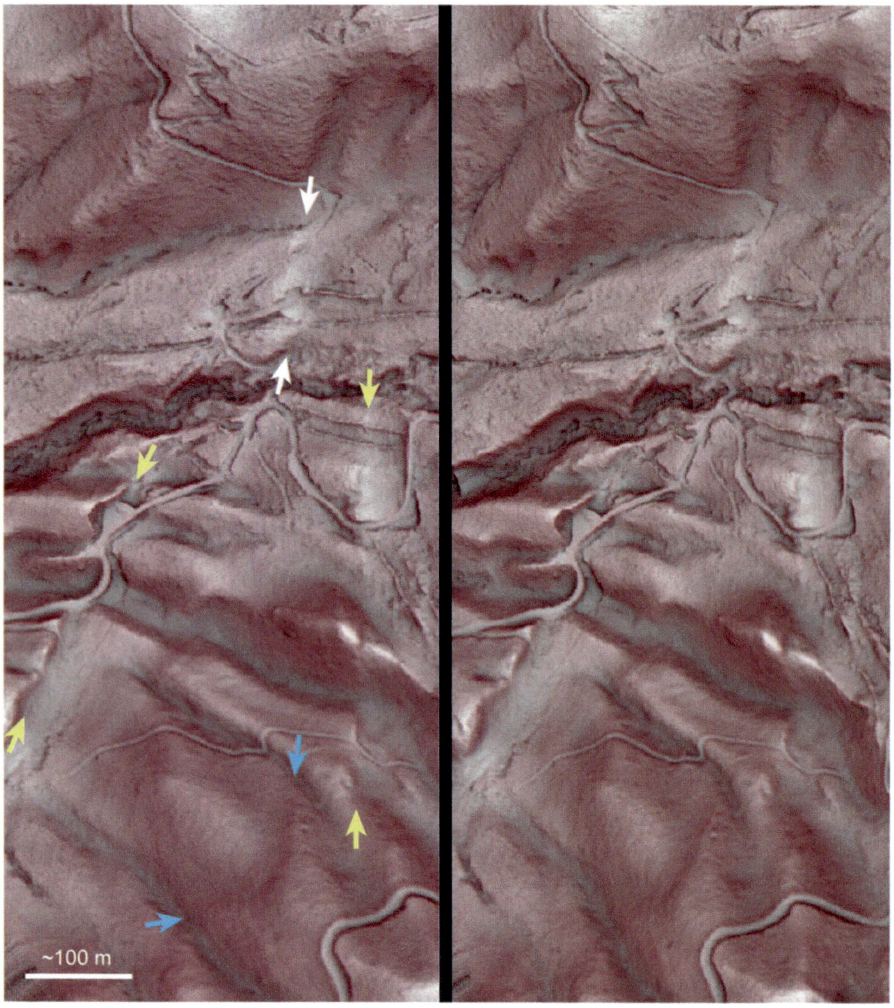

Fig. 12.37 Stereopaired morphometric protection index red relief image maps (Stereo MPI-RRIMs; Kaneda and Chiba 2019) made from the pre-earthquake 1 m grid LiDAR DEMs in the Komori Ranch. Note the presence of preexisting antithetic (uphill-facing) fault scarps (indicated by yellow arrows) as well as deformation of a fluvial terrace surface (indicated by white arrows). The northernmost part of the preexisting main normal fault scarp in the Komori Ranch is also visible and is indicated by blue arrows. The DEMs were provided by GSI. See Fig. 12.4 for location

Fig. 12.38 Photograph of a 47-cm-high uphill-facing scarp that appeared faithfully at the base of a small knoll on a ridge at Loc. 350. Looking to the west

12.7.2 Tectonic Geomorphology

As described earlier at each locality, the antithetic fault sections in the Komori Ranch are accompanied with abundant preexisting geomorphic features, in particular uphill-facing scarps that sharply cut across the mountain slope composed of middle Pleistocene pre-Aso volcanic rocks (Hoshizumi et al. 2004; Kumamoto Prefecture Geological Map Compilation Committee 2008) and similar but smaller deformation of the fluvial terrace surface that is probably abandoned in the latest Pleistocene (Fig. 12.37). Many of those features are visible on aerial photographs taken in the 1960s and 1970s because the area was already deforested then. Therefore, the main part of the most extensive antithetic fault trace (from southwest of Loc. 353 up to the north of Loc. 359) had been mapped before the 2016 earthquake as the Komori Bokujo (Ranch) fault (Research Group for Active Tectonics in Kyushu 1989). However, the antithetic traces to the east and to the west had not been identified.

Fig. 12.39 Photograph of a 45-cm-high steep ramp on a paved road formed by antithetic faulting at Loc. 351a. Looking to the west

Fig. 12.40 Photograph of a 17-cm-high antithetic fault scarp that cut across a small trail at Loc. 360. Looking to the west. The blackboard beside the trail is ~30 cm long and ~20 cm wide

12.8 Antithetic Fault Scarps in the Forest-Covered Mountains Northwest of Mt. Tawara

12.8.1 Surface Ruptures

After our initial surface rupture mapping campaign from April to May 2016, we examined the pre-earthquake 1 m grid LiDAR DEMs and found that other fresh preexisting uphill-facing scarps extend northeast–southwest to east–west in the forest-covered mountains north of the Komori Ranch or northwest of Mt. Tawara (Fig. 12.41). Almost one year after the earthquake in March 2017, we visited there and confirmed that these antithetic faults had ruptured in the 2016 earthquake.

The first evidence of the surface rupture was found at Loc. 360a on the right-bank side of an east–west trending deep valley north of the Komori Ranch (Fig. 12.4). Ground cracks grew to a clear 40–50-cm-high fault scarp at Loc. 360b. The scarp extended northeastward, making several small left stepovers, and cut across a mountain trail at Loc. 360c, where a vertical displacement was measured at 65 cm down to the southeast (Fig. 12.42). To the northeast, the surface rupture appeared along the preexisting uphill-facing scarp on which 632 m elevation is marked on the GSI topographic map (Fig. 12.4). Tree tilts were particularly evident on the upthrown side of this part of the surface rupture. From here, the co-seismic fault scarp went down the slope (Fig. 12.43), marking the maximum vertical displacement of ~80 cm along this antithetic fault trace at Loc. 360d, and reached the bottom of a small valley at Loc. 360e. Here, the co-seismic uphill-facing scarp completely cut across the valley floor, except for a small drainage gully that seemed to have been entrenched after the earthquake, forming a local dammed-up bog upstream (Fig. 12.44). Foots of living trees on the downthrown side were buried by loose peaty sediment, suggesting that the bog was also formed after the earthquake. Our pilot hand coring in the bog revealed that the peaty sediment extends down to a depth of

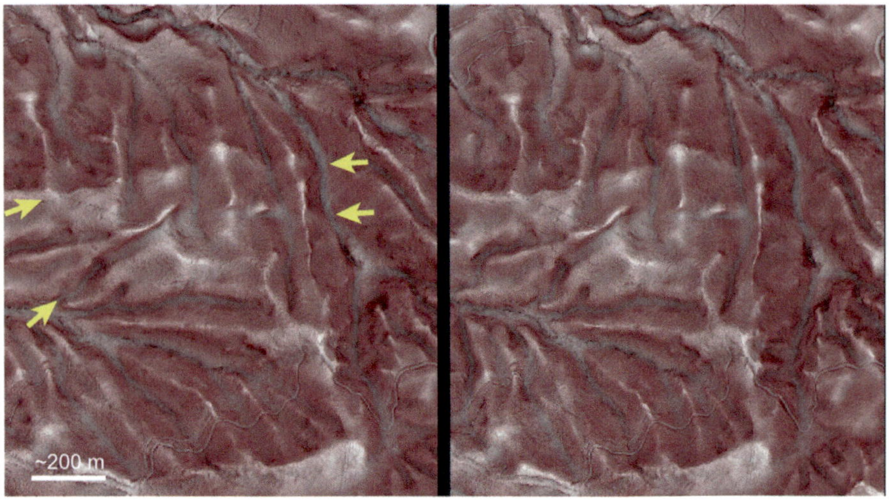

Fig. 12.41 Stereopaired morphometric protection index red relief image maps (Stereo MPI-RRIMs; Kaneda and Chiba 2019) made from the pre-earthquake 1 m grid LiDAR DEMs in the mountains north of the Komori Ranch. Yellow arrows indicate preexisting antithetic (uphill-facing) fault traces. The DEMs were provided by GSI. See Fig. 12.4 for location

Fig. 12.42 Photograph of a 65-cm-high antithetic fault scarp that cut across a small mountain trail at Loc. 360c. Looking to the west

Fig. 12.43 Photograph of a fault scarp that extended down a slope southwest of Loc. 360d. Looking to the northeast. The height of the scarp was ~80 cm

Fig. 12.44 Photograph of a fault scarp (indicated by red arrows) that extended across the bottom of a small valley at Loc. 360e. Looking to the southwest. A small dammed-up bog is in the foreground. Living trees on the downthrown side were buried by loose peaty sediment, suggesting that the bog was formed after the 2016 earthquake

Fig. 12.45 Photograph of a fault scarp (indicated by red arrows) that cut across a small valley at Loc. 360g. Looking to the east. The vertical displacement appears to be no greater than a few tens of centimeters

~90 cm and is underlain by brown to yellowish-brown clayey silt, which is interpreted to be eolian material that covered the pre-earthquake ground surface.

The fault scarp hiked up a mountain slope from the valley bottom and changed its strike from northeast–southwest to nearly east–west. At the ridge crest to the east, a 70-cm-high scarp appeared in the middle of the preexisting degraded scarp at Loc. 360f. After making small right stepovers, the surface rupture appeared in the middle of another preexisting scarp located slightly to the south of the aforementioned preexisting scarp in a right-stepping position. The surface rupture further extended down to the east and cut across a small valley at Loc. 360g, where the height of the co-seismic scarp decreased to less than a few tens of centimeters (Fig. 12.45). The fault rupture again hiked up a mountain slope, making right stepovers, and reached the ridge crest at Loc. 306 h. Interestingly, the surface rupture appeared on a small knoll of the ridge rather than on a small saddle to the south. The surface rupture soon disappeared on the slope to the east, but ~150 m further to the east, we found that a small uphill-facing scarp less than 10 cm high extended for a length of ~100 m from Locs. 360i to 360j (Fig. 12.4).

About 150 m to the north of the easternmost part of the antithetic fault trace mentioned earlier, we found another antithetic surface rupture at Loc. 360k that extended for a length of at least ~100 m. Here, the uphill-facing scarp obliquely cut across a forestry dirt road, where the vertical displacement was measured at ~55 cm down to the south (Fig. 12.46). The pre-earthquake LiDAR DEMs suggest that this antithetic fault may extend further to the east (Fig. 12.41), but we did not check if this part of the fault also ruptured in 2016.

In addition, a small local graben was formed between the antithetic fault traces mentioned earlier and the main normal fault rupture. Two subparallel, oppositely facing fault scarps appeared at Locs. 360l and 360m (Fig. 12.4), forming an ~20-m-wide local graben. Although both scarps were only ~10 cm high (Figs. 12.47 and 12.48), the co-seismic graben coincided with a preexisting small saddle on a ridge.

Fig. 12.46 Photograph of a fault scarp (indicated by red arrows) that obliquely cut across a forestry dirt road at Loc. 360k. Looking to the west. The vertical displacement is ~55 cm

Fig. 12.47 Photograph of an ~10-cm-high uphill-facing fault scarp at Loc. 360l. Looking to the northwest. The scarp bounds the northwestern margin of an ~20-m-wide local graben formed during the earthquake. Note a preexisting small knoll behind the surface rupture that suggests repetition of similar graben formation in the past

Fig. 12.48 Photograph of an ~10-cm-high fault scarp at Loc. 360m. Looking to the southwest. The scarp bounds the southeastern margin of an ~20-m-wide local graben formed during the earthquake

12.8.2 Tectonic Geomorphology

The surface rupture of this section cut across the mountain slope composed of middle Pleistocene pre-Aso volcanic rocks and partially 270–120 ka Aso-1–3 pyroclastic flow deposits (Hoshizumi et al. 2004; Kumamoto Prefecture Geological Map Compilation Committee 2008). With aerial photographs taken in the 1960s and 1970s, only some saddles can be identified along the section, and thus no active fault had been mapped before the 2016 earthquake (Research Group for Active Tectonics in Kyushu 1989). As mentioned earlier, we noticed the presence of the preexisting upfill-facing scarps that sharply cut across the mountain slopes when we were examining the pre-earthquake 1 m grid LiDAR DEMs (Fig. 12.41). The features coincide with saddles visible on aerial photographs, but their freshness can only be identified with the aid of the LiDAR survey.

12.9 Concluding Remarks

In this chapter, we described great details of the surface ruptures along and around the Idenokuchi fault, a fault interpreted to accommodate a normal-slip component of the 2016 Kumamoto earthquake (Toda et al. 2016). Almost all the surface ruptures were associated with cumulative tectonic-geomorphic signals, and therefore, those ruptures could have been predicted before the earthquake. Nonetheless, most of the northern half of the Idenokuchi fault (subsections 3–7; Fig. 12.1) had not been mapped on preexisting active fault maps (Research Group for Active Tectonics in Kyushu 1989; Ikeda et al. 2001). In particular, the largest vertical displacement associated with the earthquake (as large as $\sim$1.9 m) occurred along the unknown active fault trace (subsection 4), posing a big problem in assessing surface faulting hazards for important facilities.

We argue, however, that careful examination of proper topographic materials with proper consideration of faulting mechanics can well predict those surface ruptures. The key is to interpret tectonic-geomorphic features with keeping in mind the concept and possibility of slip partitioning and antithetic faulting. On the active fault map of Research Group for Active Tectonics in Kyushu (1989), only a short uphill-facing (antithetic) fault scarp had been mapped as the active Komori Bokujo (Ranch) fault (subsection 6), but with no "master" synthetic fault for it. Given that the right-lateral Futagawa fault to the northwest continues further to the northeast (Fig. 5.1), it would not be difficult to assume the presence of the active synthetic fault to the east of the Komori Bokujo fault along the western foot of Mt. Tawara (subsection 4), where the largest vertical displacement was observed in 2016.

Another key factor for proper active fault mapping is of course to use high-resolution LiDAR DEMs in interpreting tectonic-geomorphic features. Forest canopies mostly cover ground surfaces along the Idenokuchi fault, and therefore, it would be quite difficult to identify robust geomorphic evidence for active faulting with aerial photographs alone. With hindsight, LiDAR surveys had been conducted back in 2009, but the Kumamoto earthquake occurred before these bare-earth topographic data are to be utilized for active fault mapping. High-resolution LiDAR DEMs already cover more than half of entire Japan. More LiDAR surveys that cover the rest of the country and country-wide active fault mapping that fully and adequately uses LiDAR data are urgent tasks for a more reliable hazard assessment of seismic shaking and surface faulting.

Acknowledgements We are grateful to Geospatial Information Authority of Japan for providing airborne LiDAR data. We thank Naoya Takahashi, Tomoki Tanaka, Shintaro Kashihara, and Masahiro Gomi for their field assistance. Our gratitude also goes to Takashi Nakata for providing the information of possible surface faulting around Loc. 313a. The MPI-RRIMs used in this study are modified from the patent-protected DEM visualization of Asia Air Survey Co., Ltd. (Japanese Patent No.3670274 and No.4272146). Our fieldwork was supported by The Ministry of Education, Culture, Sports, Science and Technology (MEXT) KAKENHI (Grant Numbers 16H06298 and 16H03112), Tohoku University, and Geospatial Information Authority of Japan.

References

Hoshizumi H, Ozaki M, Miyazaki K, Matsuura H, Toshimitsu S, Uto K, Uchiumi S, Komazawa M, Hiroshima T, Sudo S (2004) Geological map of Japan 1:200,000, Kumamoto. Geological Survey of Japan, AIST, Tsukuba

Ikeda Y, Chida N, Nakata T, Kaneda H, Tajikara M, Takazawa S (2001) 1:25000 Active Fault Map in Urban Area, Kumamoto. Geospatial Information Authority of Japan

Kaneda H, Chiba T (2019) Stereopaired morphometric protection index red relief image maps (Stereo MPI-RRIMs): effective visualization of high-resolution digital elevation models for interpreting and mapping small tectonic geomorphic features. Bull Seismol Soc Am 109:2370–2391

Kumamoto Prefecture Geological Map Compilation Committee (2008) Geological Map of the Kumamoto Prefecture (1:100,000). Kumamoto Prefecture Geotechnical Consultants Association, Kumamoto

Research Group for Active Tectonics in Kyushu (ed) (1989) Active tectonics in Kyushu. University of Tokyo Press, Tokyo

Toda S, Kaneda H, Okada S, Ishimura D, Mildon Z K (2016) Slip-partitioned surface ruptures for the Mw 7.0 16 April 2016 Kumamoto, Japan, earthquake. Earth Planets Space 68:188

Surface Ruptures in the Northwestern Part of the Inner Aso Caldera

13

Takashi Nakata, Kei Tanaka, Mitsuhisa Watanabe, Satoshi Ishiguro, Takashi Kumamoto, Yasuhiro Suzuki, Hideaki Goto, Daishi Takenami, Hikaru Moriki, Shunto Tsumura, and Keita Takada

Abstract

In the northwestern part of the inner Aso caldera, distinctive fault rupture zones appeared at the time of the mainshock of the Kumamoto earthquake sequence, separated from the northern end of the Futagawa fault. They extend NE-SW for totally about 8 km in the seismically active Futae-toge zone of normal faulting (Sudo and Ikebe 2001) and the certain aftershocks of the earthquake were distributed in the fault zone. They consist of several parallel and sub-parallel sharp scarps by normal faulting as observed after the 1987 Edgecumbe earthquake in New Zealand. Along the fault traces, paddy fields were displaced vertically up to 2 m along with up to ~30-cm right-lateral horizontal displacement in places. The rupture zones develop on the paddy fields generally dipping less than 1.0 degrees on the fluvial surfaces from the central cones of the Aso volcano. It is noteworthy that the fault ruptures are distributed in the

Supplementary Information
The online version contains supplementary material available at (https://doi.org/10.1007/978-981-19-1150-7_13).
Note
The supplementary material of this book contains geographical information system (GIS) and spreadsheet files of the surface ruptures associated with the mainshock of the 2016 Kumamoto earthquake sequence

T. Nakata (✉)
Department of Geography, Hiroshima University, 1-2-3, Kagamiyama, Higashi-Hiroshima, Hiroshima, 739-8522, Japan
e-mail: tnakata@hiroshima-u.ac.jp

K. Tanaka
College of Economics, Nihon University, 1-3-2, Kanda-Misaki-cho, Chiyoda-ku, Tokyo, 101-8360, Japan
e-mail: tanaka.kei@nihon-u.ac.jp

M. Watanabe
Faculty of Sociology, Toyo University, 5-28-20, Hakusan, Bunkyo-ku, Tokyo, 112-8606, Japan
e-mail: manqq@toyo.jp

S. Ishiguro
Department of Humanities and Social Sciences, Faculty of Law and Letters, Ehime University, 3, Bunkyo-cho, Matsuyama, Ehime, 790-8577, Japan
e-mail: ishiguro.satoshi.uv@ehime-u.ac.jp

T. Kumamoto
Graduate School of Natural Science and Technology, Okayama University, 3-1-1, Tsushimanaka, Kita-ku, Okayama, Okayama, 700-8530, Japan
e-mail: tkuma@cc.okayama-u.ac.jp

Y. Suzuki
Disaster Mitigation Research Center, Nagoya University, Furo-cho, Chikusa-ku, Nagoya, Aichi, 464-8601, Japan
e-mail: ysz@nagoya-u.jp

H. Goto
Graduate School of Humanities and Social Sciences, Hiroshima University, 1-2-3, Kagamiyama, Higashi-Hiroshima, Hiroshima, 739-8522, Japan
e-mail: hgoto@hiroshima-u.ac.jp

D. Takenami
Shizuoka Local Meteorological Office, 2-1-5, Magarikane, Suruga-ku, Shizuoka, 422-8006, Japan
e-mail: p0xb3meh@s.okayama-u.ac.jp

H. Moriki
Nuclear Regulation Agency, 1-9-9, Roppongi, Minato-ku, Tokyo, 106-8450, Japan
e-mail: pyyp68v0@s.okayama-u.ac.jp

S. Tsumura
Department of Geography, Faculty of Letters, Hiroshima University, 1-2-3, Kagamiyama, Higashi-Hiroshima, Hiroshima, 739-8522, Japan

K. Takada
Fukken Co., Ltd., 3-8-15, Iwamoto-cho, Chiyoda-ku, Tokyo, 101-0032, Japan
e-mail: takada@fukken.co.jp

Y. Kumahara et al. (eds.), *Surface Ruptures Associated with the 2016 Kumamoto Earthquake Sequence in Southwest Japan*, Advances in Geological Science, https://doi.org/10.1007/978-981-19-1150-7_13

area with seismic intensity less than the JMA scale 5+. From the characteristics of the surface fault ruptures, we conclude that the ruptures were triggered by normal faulting with right-lateral slip during the mainshock, but not simple lateral spreading caused by shaking.

Keywords

Normal fault ruptures • Aso caldera • Drone photogrammetry

Distinctive fault ruptures appeared in the northwestern part of the inner Aso caldera, separated from the ruptures around Kuro-kawa River at the northern end of the Futagawa fault (Chap. 10). These ruptures appeared in a zone extending NE-SW within the Futae-toge seismic zone of normal faulting (Sudo and Ikebe 2001). These ruptures were similar to those associated with historical normal faulting earthquakes, such as the 1983 Borah Peak earthquake in Idaho, USA (Crone et al. 1987) and the 1987 Edgecumbe earthquake in New Zealand (Beanland et al. 1989) (Fig. 13.1).

In the northwestern part of the inner Aso caldera, the alluvial lowland consists mainly of fluvial surfaces extending from the central cones of the Aso volcano. Along several streams are gently sloping fan-shaped surfaces covered with black humus layers. We can also identify abandoned channels, natural levees, back marshes along the Kuro-kawa River, and talus-colluvium surfaces at the foot of the caldera wall (Fig. 13.2). Several researchers associated these ruptures with lateral spreading caused by strong ground motion (e.g., Kuroki et al. 2016). However, these ruptures were quite different from typical lateral spreading observed in severely shaken areas around Budharmora due to the January 26, 2001, Republic Day (Bhuj) earthquake of India (Nakata et al. 2001). Kim et al. (2017) interpreted that these ruptures appeared following abandoned channels of the Kuro-kawa River, and Haraguchi (2017) claimed that the ruptures appeared in the former limonite mines. However, these interpretations are not supported by the distribution of the ruptures and former mines and geomorphic classification, as shown in Fig. 13.2. We estimated the seismic intensity of the area where the surface ruptures appeared based on the damage of graveyards (Fig. 13.3). We found that the seismic ground motion in this area was much smaller than the area along the Futagawa fault. It was somewhat surprising that most of the tombstones, including long and unstable ones, were not severely affected and remained upright even in the area close to the surface ruptures, while the intensity in the surrounding areas, such as the foot of the caldera wall, was much higher (Fig. 13.3). These observations suggested that the intensity of the area was less than the JMA scale 5+. Roof tiles of houses very close to or even on the fault ruptures were not dropped at all. Besides, the surface slope of the fluvial plain is minimal (much less than 1°), as Mukoyama et al. (2016) pointed out. Mukoyama et al. (2016) analyzed surface displacements in the northwestern Aso caldera by comparing LiDAR DEM images before and after the 2016 Kumamoto earthquake and recognized the horizontal and vertical displacements in a wide area. Fujiwara et al. (2016) analyzed ALOS-2 satellite radar interferometry images. They found more than 2 m horizontal displacement in an NNW direction around Uchinomaki hot spring area and concluded that the surface ruptures were not caused tectonically but by liquefaction due to strong ground motion. Hase et al. (2010) reported that lacustrine clay and silt as thick as 70 m underlie the lowland along the Kuro-kawa River based on borehole data analysis. Tsuji et al. (2017) reported that horizontal movement took place along the unconformity between the lacustrine sediments and older fluvial sediments at a depth of 50 m based on borehole camera observation and bending of pipes of hot spring wells around Uchinomaki. They illustrated presumed normal faulting beneath the surface ruptures but vaguely suggested that lateral sliding (spreading) was triggered by strong shaking from the mainshock (Tsuji et al. 2017).

Fig. 13.1 Photograph of a typical normal faulting surface rupture. The rupture was associated with the 1987 Edgecumbe earthquake in the North Island of New Zealand. From GeoNet Web site (GNS Science 2021)

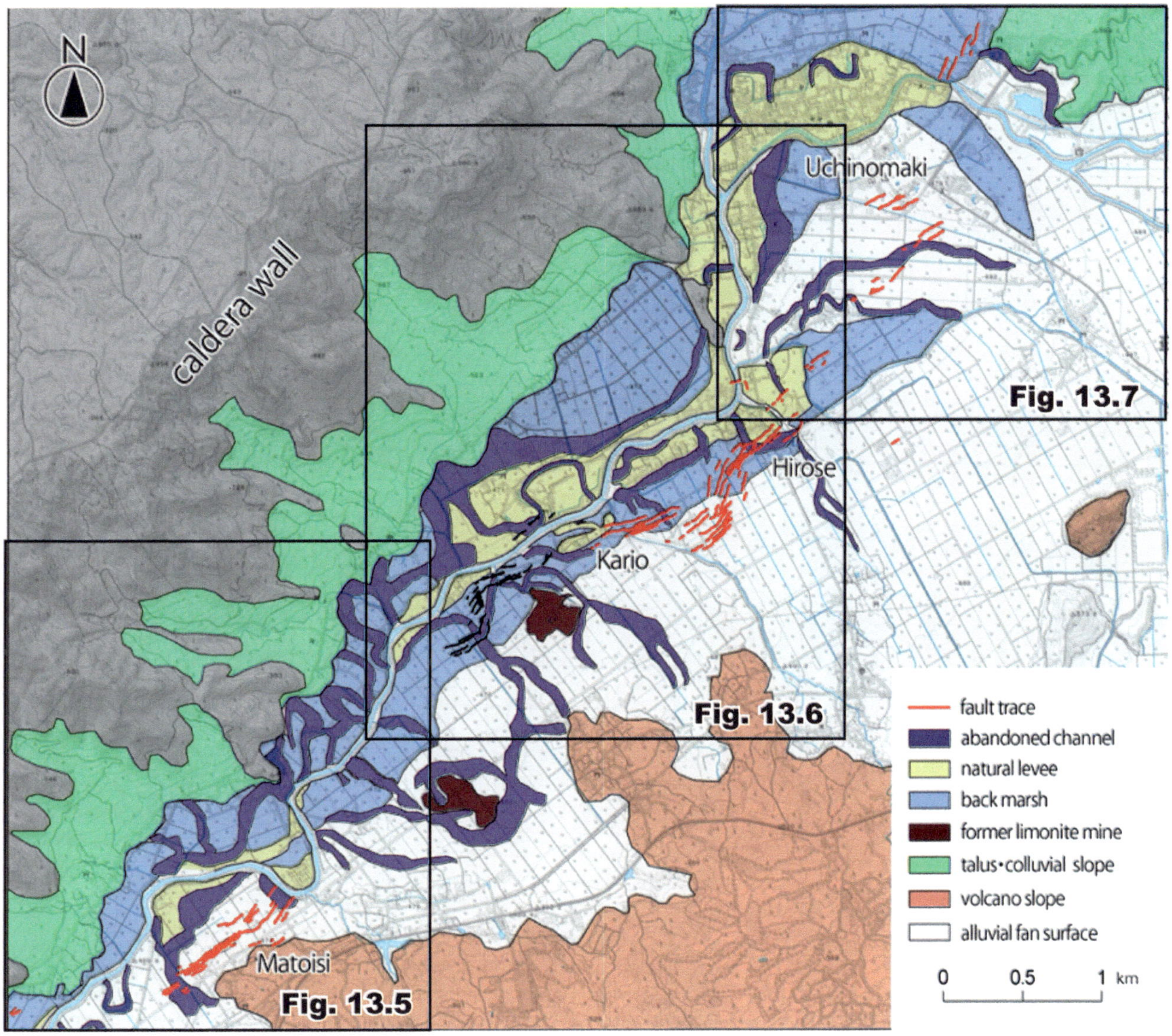

Fig. 13.2 Geomorphic classification of the northwestern part of the inner Aso caldera with the surface ruptures during the 2016 Kumamoto earthquake. The ruptures west of Kario followed abandoned channels

The occurrence of shallow small earthquakes characterizes the surface rupture zone in the northwestern Aso caldera. We plotted the epicenters of earthquakes (purple dots) that occurred before April 14, 2016 (Fig. 13.4). Sudo and Ikebe (2001) named the seismic zone the Futae-toge earthquake zone. Sudo et al. (1984) pointed out that the earthquakes in the Aso caldera were shallower than 8 km. The depth of M3.6 and M2.2 normal faulting earthquakes on June 12, 1983, was 4.5 km and 5.3 km, respectively. The 2016 surface ruptures extended within the seismic zone.

Many small aftershocks (yellow dots) between April and May 2016 formed an ENE-WSW zone beneath the fault ruptures. The southeast boundary of the seismic zone roughly coincided with the boundary of horizontal movement of the ground (Mukoyama et al. 2016) and the southeast limit of large deformation (Fujiwara et al. 2016). Shito et al. (2017) analyzed the spatial distribution of seismicity of the Kumamoto earthquake sequence. They illustrated that small earthquakes shallower than 5 km were dominant before the 2016 earthquake, but all the aftershocks were deeper than 5 km in the northwestern part of the inner Aso caldera. We interpret that the stress at the shallow crust decreased due to large near-surface coseismic slip. Therefore, we conclude that the surface ruptures resulted from normal faulting with right-lateral slip during the mainshock on April 16 and triggered surface deformation, including extensive lateral spreading.

We describe the location and geometry of the surface ruptures from west to east (Figs. 13.5, 13.6 and 13.7). The surface rupture appeared at the foot of the caldera wall in

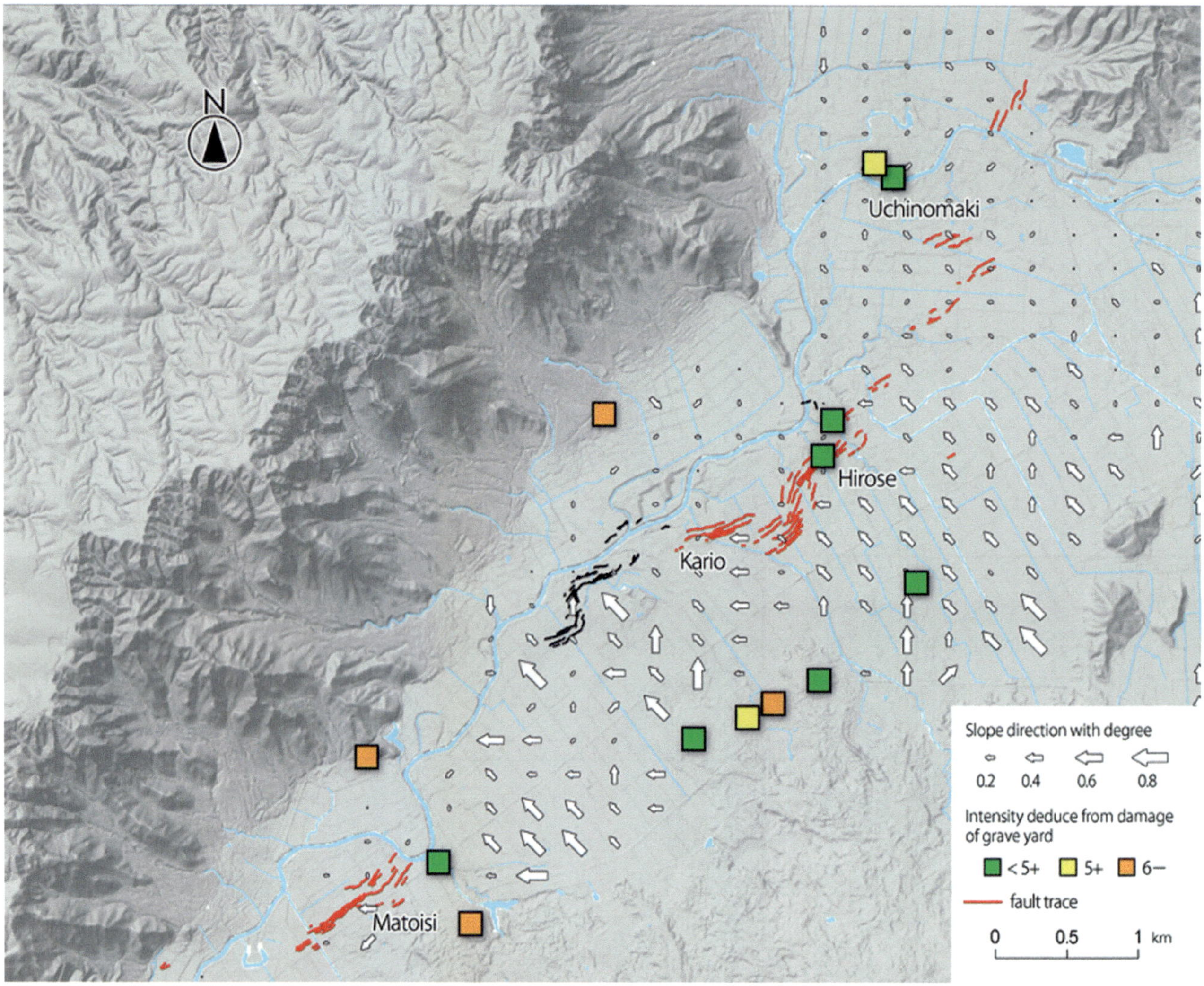

Fig. 13.3 Surface fault ruptures with presumed seismic intensity and slope direction and degree. Most of the ruptures appeared on the very gently dipping surfaces (<1°), except for west of Kario. Red lines denote surface ruptures due to normal faulting, and black lines denote surface ruptures due to surface failure. Colored squares indicate JMA seismic intensity deduced from the extent of damage to the graveyards; Green: less than 5+ with no tombstones fallen, Yellow: 5+ with few tombstones fallen, Orange: 6- with many tombstones fallen

Tonozuka as a flexure scarp on the paved road at Loc. C1 (Fig. 13.5). This rupture seemed to be located on an eastern extension of the rupture across the prefectural road above the caldera wall. Vertical offset of about 40 cm up to the south was estimated from a topographic profile obtained from total station measurement (Fig. 13.8). About 100 m east from Loc. C1, two parallel-running ruptures extending NE-SW appeared in the paddy field (Fig. 13.9). Along the 100-m-long NW rupture, the paddy field was displaced about 20 cm down to the south (Loc. C2). The displacement was about 40 cm down to the north along the SE rupture (Loc. C3). Between these ruptures was a 20-m-wide depression. The houses in Tonozuka village were severely damaged, and their roof tiles slipped down.

In Matoishi on the southeast bank of the Kuro-kawa River northeastward to the Aso-Nishi Elementary School, several ruptures formed an N60°E-trending, ~1-km-long and 200-m-wide at maximum, rupture zone on the gently dipping paddy field (Figs. 13.3 and 13.5). We flew a drone to make a detailed orthoimage and digital surface model (DSM) and made shaded-relief maps and cross-sections (Fig. 13.10). The southernmost rupture was most prominent, and the near-vertical scarp cut the paddy field for 500 m (Fig. 13.10). Fourteen profiles across the rupture zone well illustrate the lateral transformation of the fault scarps. Along the fault trace, paddy fields were displaced up to 2 m down to the north around Loc. C6 (Fig. 13.11). A narrow 20-m-wide depression was formed between the north-facing prominent scarps and sharp south-facing scarps ~40 cm high, as typically observed around Loc. C10 in Matoishi. A depression north of the rupture zone may coincide with the location of the former limonite mine (Haraguchi 2017).

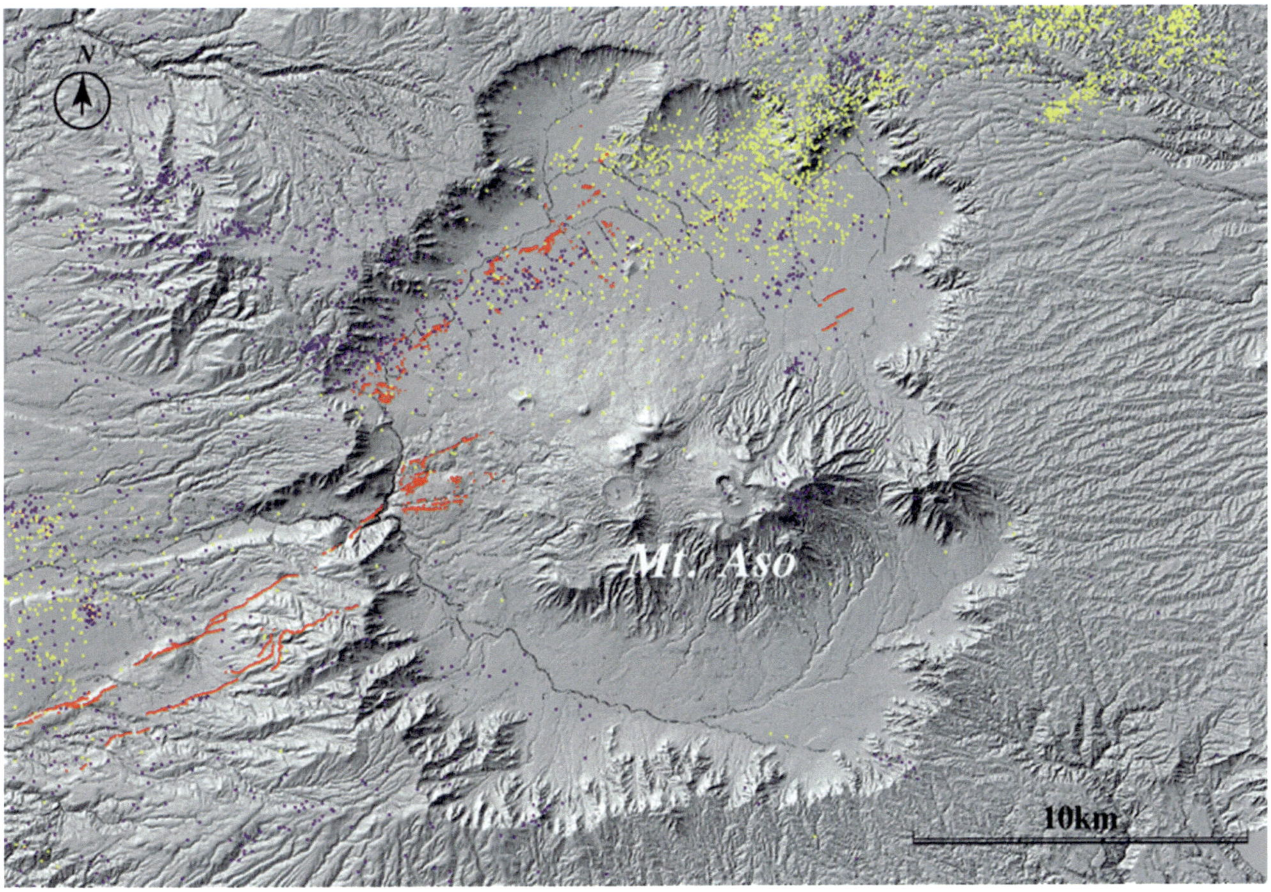

Fig. 13.4 Epicenters of earthquakes plotted by SEIS-PC (Odaka and Ishikawa 1993) using JMA data. Earthquakes greater than M = 1 and shallower than 20 km in and around the study area are shown. Yellow dots show earthquakes between April and May 2016, and purple dots between January 1996 and December 2015

Figure 13.12 shows the western part of this rupture zone. A temporary pond was formed in the depression. The rupture zone became wider eastward, and the slip was distributed to more fault traces than in the west. The eastern extension of the fault traces extended up to the west bank of the meandering Kuro-kawa River.

Between Matoishi and Kario, the surface ruptures were not recognizable in the former limonite mining area (Fig. 13.2). The surface ruptures about 500 m west of Kario mostly followed the abandoned channel of the Kuro-kawa River, and they may not be tectonic in origin. Around these ruptures, Kuroki et al. (2016) did not detect horizontal movement of the surface. We speculate that these ruptures were caused by other reasons such as liquefaction.

Around Kario, a 600-m-long and 80-m-wide rupture zone comprises three or four parallel ruptures (Fig. 13.6). We flew a drone to make a detailed orthoimage and DSM and then made shaded-relief maps and cross-sections (Fig. 13.13). The parallel surface faults were relatively straight (Fig. 13.14), and the southernmost trace exhibited the maximum slip of 130 cm up to the south at Loc. C17 on profile 9 in Fig. 13.13. In the western part of the rupture zone, between profiles 10 and 12, the slip was distributed on four or five fault traces with each slip less than 50 cm, except for 1.5 m slip up to the north on the northernmost fault trace at Loc. C13 on profile 13 (Fig. 13.15). On the contrary, in the eastern part of the rupture zone, a ~50-m-wide elongated depression extended for ~300 m between profiles 3 and 9 at Locs. C14 to C17 and along profile 10 in Fig. 13.13. The fault ruptures were associated with ~30 cm right-lateral displacement. South of the ruptures in Kario described above, there was a 100-m-wide rupture zone with more than five fault traces. This zone extended to the northeast with a broad S-shaped bend to Hirose.

Around Hirose, a distinctive elongated depression formed between parallel fault traces (Fig. 13.16). The faults on both sides of the depression showed large displacements. The vertical separation down to the south on the northern rupture was 100 cm at Loc. C18 and 120 cm at Loc. C19. On the southern rupture, the vertical separation down to the north was 170 cm at Loc. C20 and 100 cm at Loc. C22. The

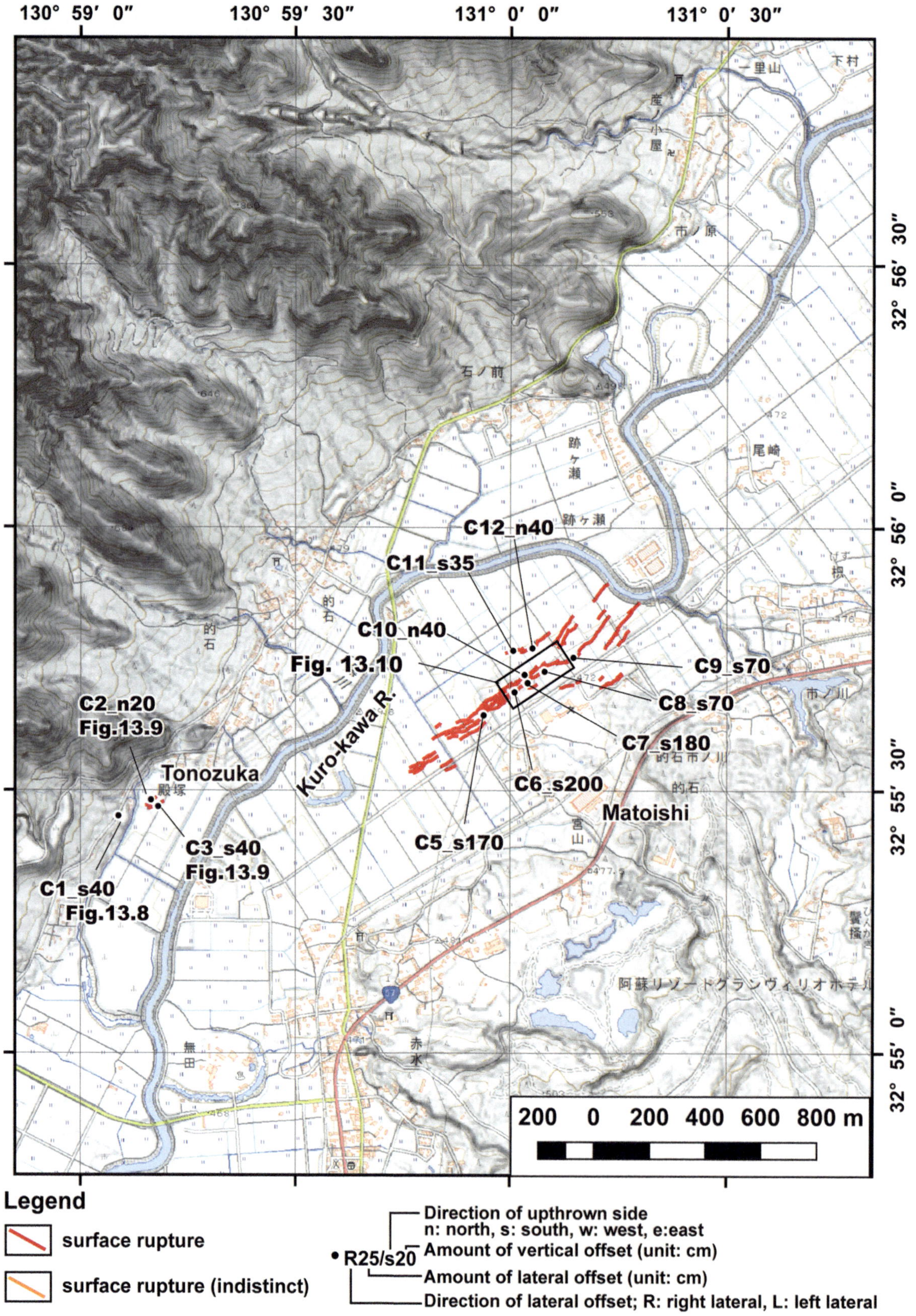

Fig. 13.5 Detailed surface rupture map of the northwestern part of the inner caldera around Matoishi, Aso City. Numerals denote strike-slip and vertical displacements at each locality. The base map is a half-transparent topographic map of Geospatial Information Authority of Japan (GSI) underlain by a slope map produced from 5 m grid digital elevation models of GSI. The location of the map is shown in Fig. 13.2

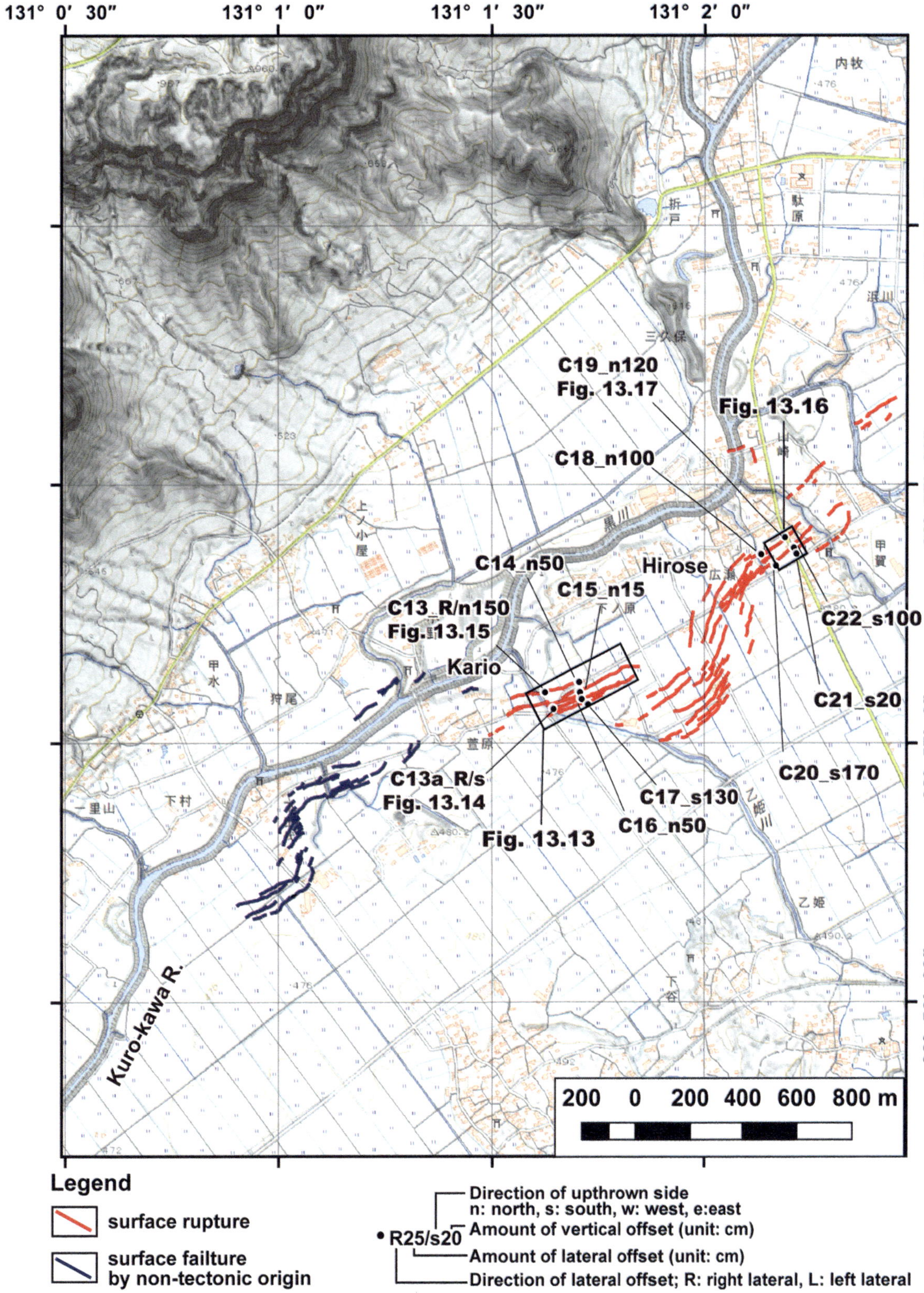

Fig. 13.6 Detailed surface rupture map of the northwestern part of the inner caldera around Kario and Hirose, Aso City. Legends are the same as Fig. 13.5. The location of the map is shown in Fig. 13.2

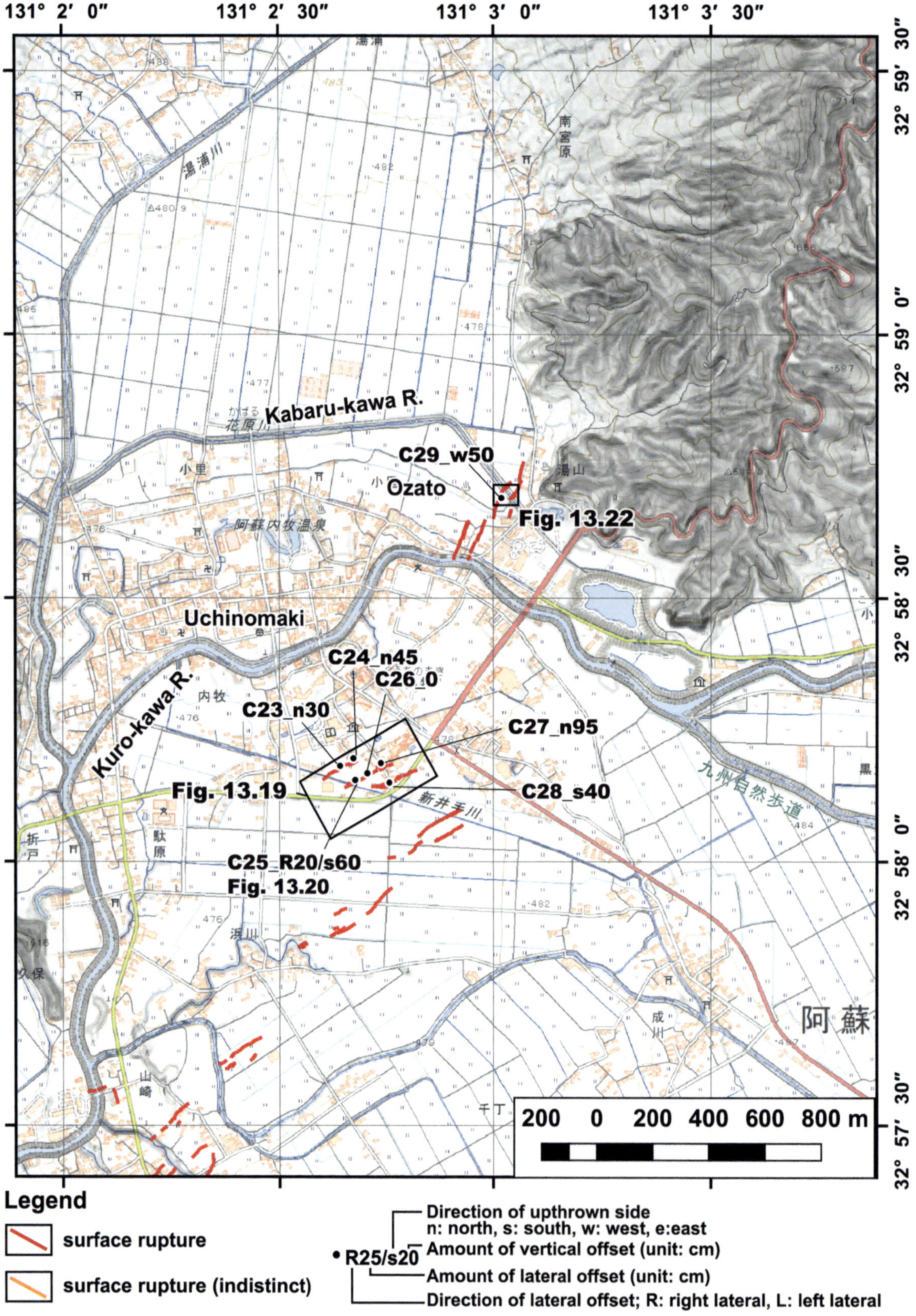

Fig. 13.7 Detailed surface rupture map of the northwestern part of the inner caldera around Uchinomaki, Aso City. Legends are the same as Fig. 13.5. The location of the map is shown in Fig. 13.2

Fig. 13.8 Photograph of the vertical displacement (40 cm up to the south) on the road at Loc. C1

Fig. 13.9 Photograph of the surface rupture across the paddy field in Tonozuka, looking southeast. The maximum vertical separation was 40 cm up to the south. A narrow depression 20 m wide appeared between this rupture and a rupture with 20 cm down to the south displacement

houses very close to the surface rupture had not been shaken hard, even on the edge of the fault scarp, because their roof tiles were not moved or dropped (Fig. 13.17). The red-roof houses in Fig. 13.16 moved ~100 cm down to the bottom of the depression. They were not damaged much, and the residents did not feel that they were brought down with the houses. We identified another significant evidence of minor shaking at a graveyard on the northern rupture (Fig. 13.18). The graveyard was faulted and displaced ~1 m vertically, but the tomb on the fault was only tilted but not destroyed. The other tombs were not affected at all by shaking, and unstable pillar stones remained upright. These observations indicate that the coseismic ground shaking was not severe, although vertical fault separation was more than 1 m. Between Hirose and Uchinomaki, we identified the NE-trending 800-m-long and 50-m-wide rupture zone (Fig. 13.7).

South of Uchinomaki, three distinctive parallel surface ruptures cut paddy field, residential area, and parking lot of an amusement store (Figs. 13.7 and 13.19). The surface ruptures were pretty straight. The length of the northern, central, and southern traces was ~150 m, ~200 m, and ~100 m, respectively. Along the northern trace, we measured 30 cm and 45 cm north-up vertical separation at Locs. C23 and C24, respectively. Along the central trace, the vertical displacement varied along the strike; south-up separation in the west (60 cm at Loc. C25, Fig. 13.20), no separation in the middle (~0 cm at Loc. C26), and north-up separation in the east (95 cm at Loc. C27, Fig. 13.21). South-up vertical separation was observed along the southern trace, such as 40 cm on the road at Loc. C28. We measured right-lateral offset less than 20 cm of the white parking lines of the parking lot (Fig. 13.20), suggesting that these surface ruptures were normal faults with a right-lateral slip component. The northeasternmost surface ruptures appeared north of Uchinomaki as NE-trending parallel traces toward the foot of the caldera wall, forming a 50-m-wide shallow depression at Ozato (Fig. 13.22). Vertical separation of the paddy field by the western rupture was ~50 cm down to the east at Loc. C29.

From the characteristics of the surface ruptures formed during the mainshock of the 2016 Kumamoto earthquake, we conclude that the ruptures resulted from normal faulting with right-lateral slip associated with shallow earthquakes in the northwestern part of the inner Aso caldera.

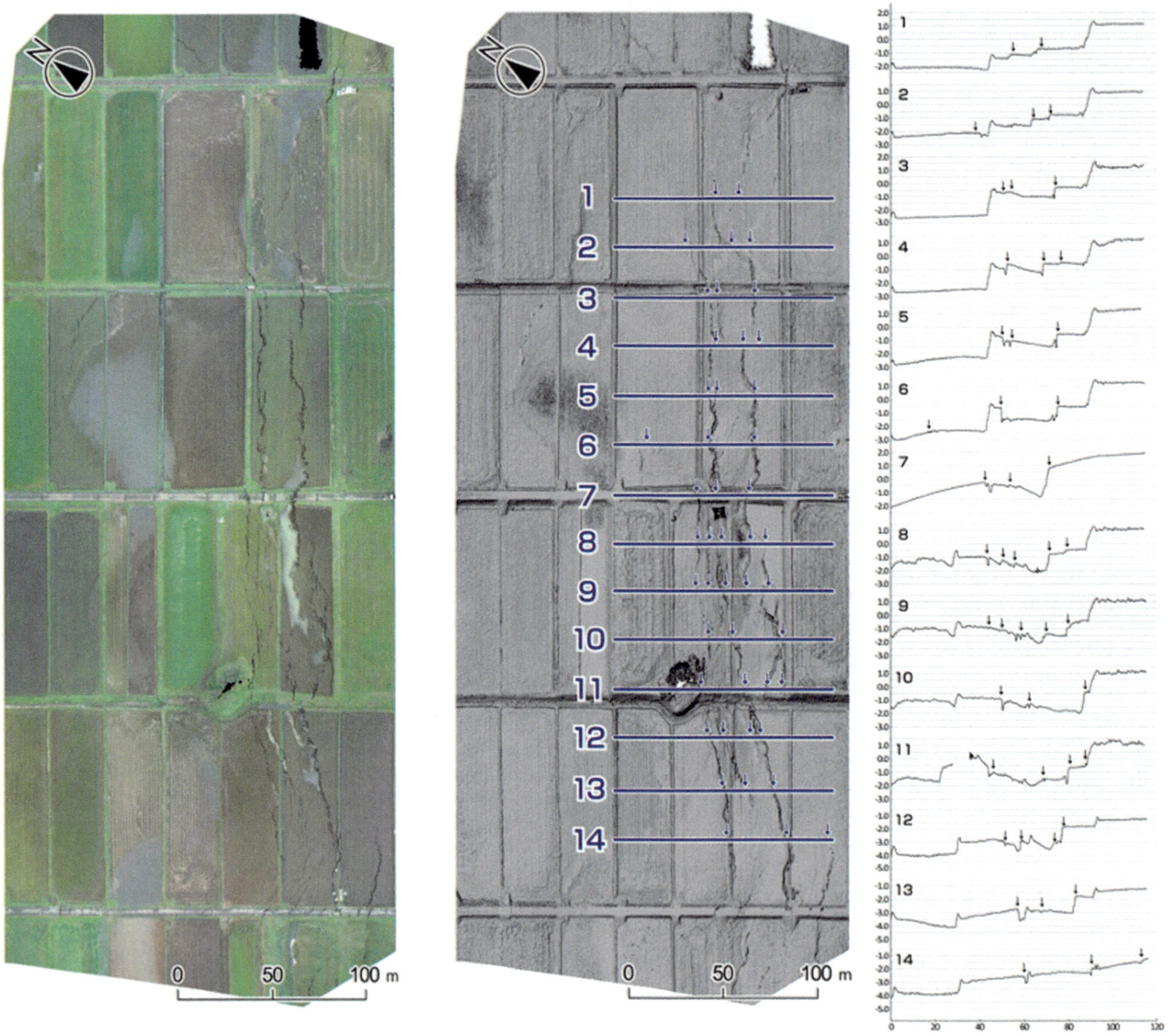

Fig. 13.10 Compiled drone images of the surface ruptures around Matoishi. Left: Ortho image, Right: Gray-scale shaded-relief image with topographic profiles every 25 m across the fault ruptures. Small arrows on the profiles indicate the location of the fault ruptures. The location of the images is shown in Fig. 13.5

Fig. 13.11 Photograph of the distinctive fault rupture ~ 2 m high in front of the Aso-Nishi Elementary School, looking south. The topographic profile along the displaced road is shown on Profile 7 in Fig. 13.10

Fig. 13.12 Photograph of the surface rupture around Profile 11 in Fig. 13.10. The tree-covered low mound seen in the distance existed before 2016, and it was relatively uplifted by normal faulting at the time of the earthquake, suggesting cumulative fault activities

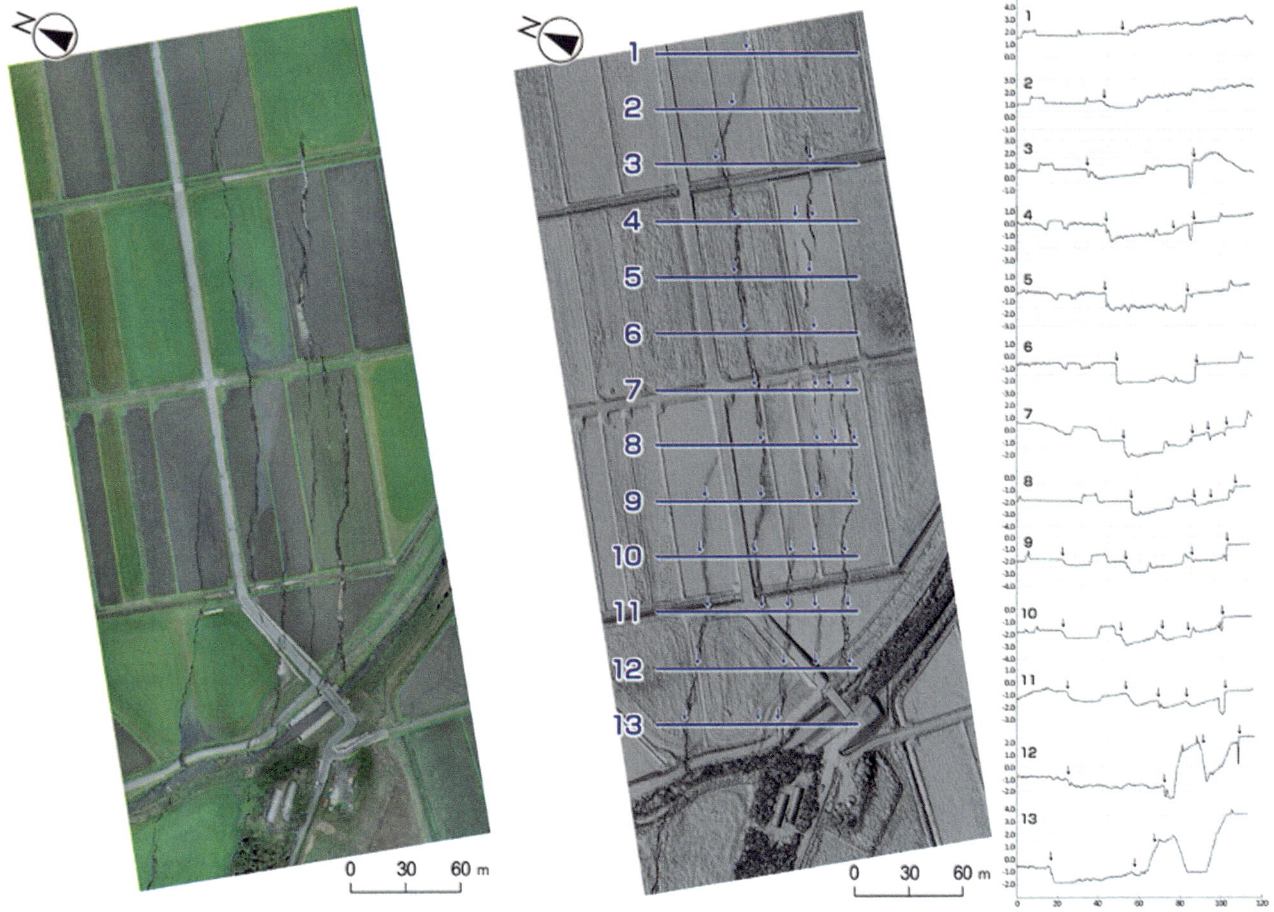

Fig. 13.13 Compiled drone images of the surface ruptures around Kario. Left: Ortho image, Right: Gray-scale shaded-relief image with profiles every 25 m across the fault ruptures. Small arrows on the profiles indicate the location of the fault ruptures. The location of the images is shown in Fig. 13.6

Fig. 13.14 Photograph of the rather straight fault traces down to the north of the southernmost fault in Fig. 13.13. The right-lateral displacement of the irrigation channel is apparent

Fig. 13.15 Photograph of the right-lateral slip of the irrigation channel and down to the south displacement along the northernmost fault on profile 13 in Fig. 13.13

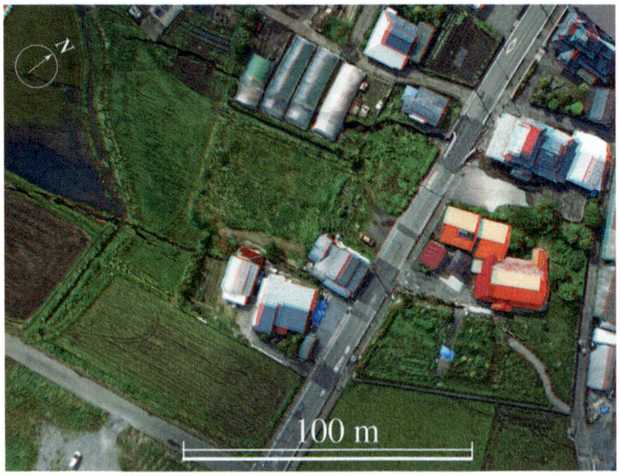

Fig. 13.16 Anaglyph from drone images at Hirose. Repaved zones of the main road were faulted, and a ~50-m-wide depression appeared between the parallel faults. Red-roof houses were carried down to the floor of the depression without severe shaking damage

Fig. 13.17 Photograph of the road displaced by 120 cm up to the north at Hirose. The house on the edge of the fault escaped from structural damage, suggesting that severe shaking was not associated with faulting. In other words, gravitational lateral spreading due to severe shaking was not the cause of the fault scarp

Fig. 13.18 Photograph of the graveyard on the fault. Even unstable pillar stones of the tombs did not fall, and the houses behind also escaped from damage. The tilted tomb on the warped slope by faulting indicates that seismic ground motion was not strong enough to cause gravitational deformation. The surface rupture appeared along the edge of the paddy field

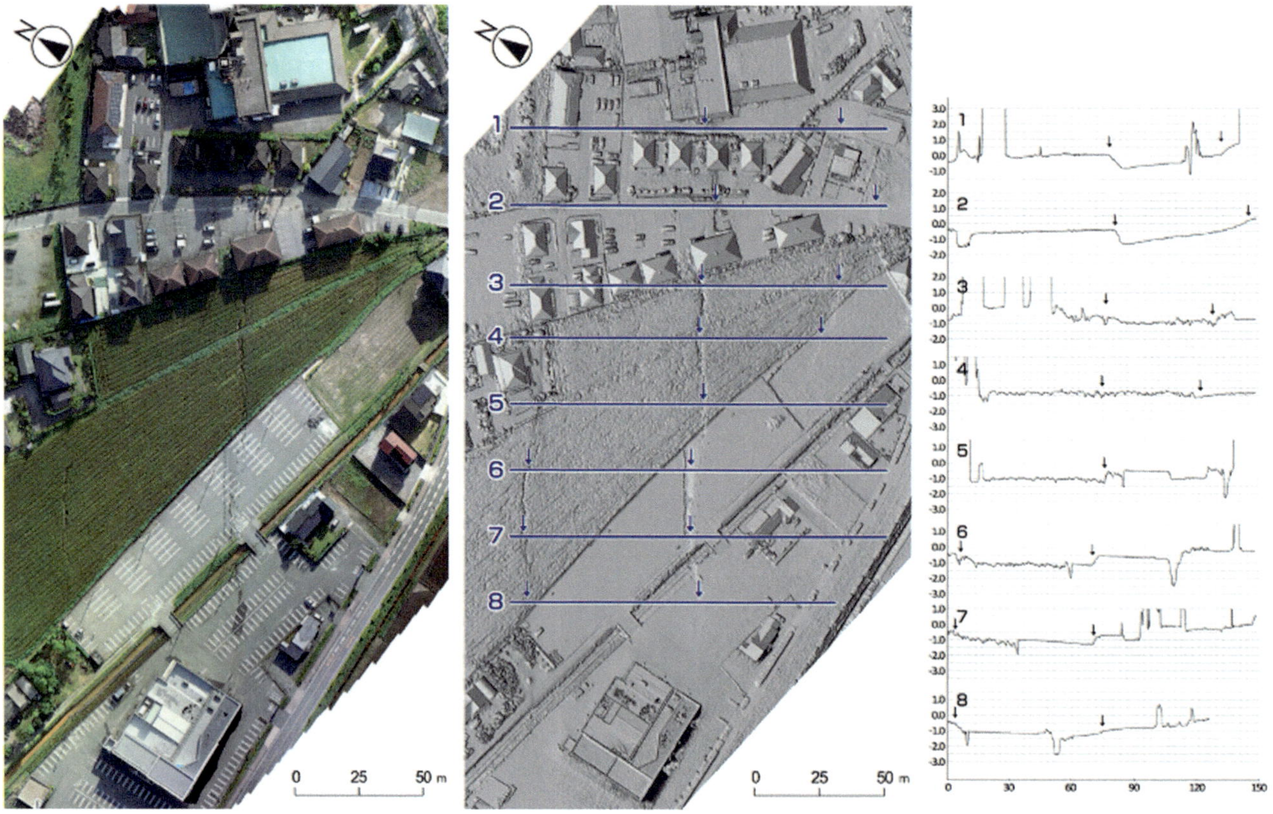

Fig. 13.19 Compiled drone images of the surface ruptures around Uchinomaki. Left: Ortho image, Right: Gray-scale shaded-relief image with profiles every 25 m across the fault ruptures. Small arrows on the profiles indicate the location of fault ruptures. The location of the images is shown in Fig. 13.7

Fig. 13.20 Photograph of the central rupture south of Uchinomaki that cut across the parking lot with 60 cm separation up to the south. The white parking lines were offset right-laterally less than 20 cm. An apartment house on the fault trace was tilted southward due to normal faulting

Fig. 13.21 Photograph of the 95 cm displacement of the road in the eastern part of the central rupture south of Uchinomaki. A house on the fault was tilted. The fault trace crossed a hotel building behind. The building tilted slightly, but distinctive cracks were not detected from outside

Fig. 13.22 Ortho image compiled from drone photographs taken on August 8, 2016, at Ozato. Arrows indicate the location of surface ruptures. The location of the image is shown in Fig. 13.7

References

Beanland S, Berryman K, Blick G (1989) Geological investigation of the 1987 Edgecumbe earthquake, New Zealand. New Zealand Journal of Geology and Geophysics 32:73–91

Crone A, Machette WN, Bonilla MG, Lienkaemper JJ, Pierce KL, Scott WE, Bucknam RC (1987) Surface faulting accompanying the Borah Peak earthquake and segmentation of Lost River fault, central Idaho. Bull. Seis. Soc. America 77(3):739–770

Fujiwara S, Yarai H, Kobayashi T, Morishita Y, Nakano T, Une H, Miyahara B, Nakai H, Miura Y, Ueshiba H, Kakiage Y (2016) Small-displacement linear surface ruptures of the 2016 Kumamoto earthquake sequence detected by ALOS-2 SAR interferometry. Earth, Planets and Space 68:160

GNS Science (2021) GeoNet, https://www.geonet.org.nz/earthquake/story/04228 Accessed 24 August 2021.

Haraguchi T (2017) Classification of the relationship between the damage distribution and surface cracks caused by the 2016 Kumamoto earthquake. In: Abstracts of the Japan Geoscience Union Meeting, Makuhari Messe, Chiba, 20–25 May 2017

Hase Y, Miyabuchi Y, Haruta N, Sasaki N, Yumoto T (2010) Change of sedimentary facies and topographic process after the late period of the last glacial age of Asodani in the northern area of Aso caldera in central Kyushu, Japan. Bulletin of Goshoura Cretaceous Museum 11:1–10

Kuroki T, Iguchi T, Chida K, Kato Y, Omata M, Yorida K, Takami T, Tajika J, Tamura H, Nishiyama K, Yada J, Mouyama S, Nishimura T, Sakaguchi K, Suetake S, Hashimoto S, Teraguchi K, Yano K, Yamamoto S, Usui T, Yano H, Sato T, Kawai T, Ishibashi A (2016) Horst Graben and its surrounding terrain displacement that appeared in the Aso caldera. Japan Society of Engineering Geology. http://www.jseg.or.jp/00-main/pdf/20160716_Aso.pdf. Accessed 27 July 2021

Kim H-Y, Nagai M, Haraguchi T (2017) Geomorphological and geological factors of surface cracks in Aso-caldera in association with the 2016 Kumamoto earthquake. In: Abstracts of the Japan Geoscience Union Meeting 2017, Makuhari Messe, Chiba, 20–25 May 2017

Mukoyama S, Sato T, Takami T, Nishimura T (2016) Estimation of ground displacements around Aso-Caldera caused by the 2016 Kumamoto Earthquake, from the geomorphic image analysis of temporal LiDAR DEMs. Japan Society of Eng Geology. http://www.jseg.or.jp/00-main/pdf/20160822_kumamoto_rep.pdf Accessed 27 July 2021

Nakata T, Yoshioka T, Sato H, Imaizumi T, Malik JN, Philip G, Maharajan AK, Karanth RV (2001) Extensive surface deformation around Bhudhamora associated with the January 26, 2001 Republic Day (Bhuj) earthquake of India. Active Fault Res 20:127–136

Odaka T, Ishikawa Y (1993) Application of "SEIS-PC" to the 3D visualization of hypocenter distribution. Geoinformatics 4:397–402

Shito A, Matsumoto S, Shimizu H, Ohkura T, Takahashi H, Sakai S, Okada T, Miyamachi H, Kosuga M, Maeda Y, Yoshimi M, Asano Y, Okubo M (2017) Seismic velocity structure in the source region of the 2016 Kumamoto earthquake sequence, Japan. Geophys Res Lett 44:7766–7772

Sudo Y, Yamada T, Masuda H (1984) Seismic activity and focal mechanism around Aso Caldera. DPRI Annuals Kyoto Univ. B-1 27:35–44

Sudo Y, Ikebe S (2001) New active fault described in Aso Caldera and seismic activity. DPRI Annuals Kyoto Univ B-1 44:345–352

Tsuji T, Ishibashi J, Ishitsuka K, Kamata R (2017) Horizontal sliding of kilometre-scale hot spring area during the 2016 Kumamoto earthquake. Sci Rep 7:42947. https://doi.org/10.1038/srep42947

Surface Ruptures in the Northeastern Part of the Inner Aso Caldera

14

Daisuke Ishimura and Shinji Toda

Abstract

A large number of distributed surface ruptures associated with the mainshock of the 2016 Kumamoto earthquake sequence appeared apart from the primary surface ruptures. In the northeastern part of the inner Aso caldera, small displacements were detected as the discontinuities of fringes in InSAR images. This chapter reports the result of the field examination of these discontinuities. We found two surface rupture traces with northeast-southwest strikes and measured vertical and right-lateral offsets on the both ruptures based on the damaged paved roads and offset of block walls.

Keywords

Distributed surface ruptures • Aso caldera • InSAR • Right-lateral offset

Supplementary Information
The online version contains supplementary material available at (https://doi.org/10.1007/978-981-19-1150-7_14).
Note
The supplementary material of this book contains geographical information system (GIS) and spreadsheet files of the surface ruptures associated with the mainshock of the 2016 Kumamoto earthquake sequence

D. Ishimura (✉)
Department of Geography, Tokyo Metropolitan University, 1-1, Minami-Osawa, Hachioji, Tokyo, 192-0397, Japan
e-mail: ishimura@tmu.ac.jp

S. Toda
International Research Institute of Disaster Science (IRIDeS), Tohoku University, 468-1, Aoba, Aramaki, Aoba-ku, Sendai, Miyagi, 980-0845, Japan
e-mail: toda@irides.tohoku.ac.jp

InSAR revealed not only large and wide deformation associated with the mainshock but also small displacements apart from the primary surface rupture traces (e.g., Fujiwara et al. 2016). These small displacements were detected based on apparent discontinuities of fringes in InSAR images. In the northeastern part of the inner caldera, we identified two linear discontinuities in an InSAR image and conducted the first field survey on 9 May 2016. We found two strands (hereafter, north and south strands) of small surface ruptures just on the discontinuities. Both strands trended northeast with right-lateral slip (Fig. 14.1). These surface ruptures appeared on the alluvial fan surface at the foot of the Aso volcano, where no active fault traces were previously mapped.

The ~1-km-long north strand showed a slightly sinuous trace (Fig. 14.1) with a small amount of right-lateral offset (<5 cm) and vertical offset (<10 cm) accompanied with open cracks. Around the northeast end (Figs. 14.2, 14.3, 14.4, and 14.5), southeast-side-down vertical displacement (amount <5 cm) was traceable from Loc. 423b to Loc. 423. Immediately southwest of Loc. 423 (Fig. 14.5), Ishimura et al. (2017) excavated a 1.5 m-deep pit to examine the subsurface fault structures and found almost vertical faults with a flower structure. They also recognized that the vertical offset reached 15–20 cm at 1.5 m depth and the displacement tapered to the ground surface. Surface rupture traces on the central part of the north strand were unclear, and only open cracks on paved roads were identified at Locs. 422b (Fig. 14.6) and 422a (Fig. 14.7). Around the southwest end (Figs. 14.8, 14.9, 14.10, and 14.11), northwest-side-down vertical offset (amount <10 cm) was traceable from Loc. 420 to Loc. 421. At Loc. 422, a 5-cm right-lateral offset was measured using a block wall as a piercing point (Fig. 14.8).

On the ~2-km-long south strand, evident surface ruptures were not identified, whereas broad deformation showing extension (e.g., open cracks on paved roads) was identified (Figs. 14.12, 14.13, 14.15, and 14.16). Open cracks on paved road distributed within a 10–20-m-wide

Y. Kumahara et al. (eds.), *Surface Ruptures Associated with the 2016 Kumamoto Earthquake Sequence in Southwest Japan*, Advances in Geological Science, https://doi.org/10.1007/978-981-19-1150-7_14

Fig. 14.1 Detailed surface rupture map of the northeastern part of inner Aso caldera. See Fig. 5.2 for location of the map. Strike-slip and vertical displacements at each locality are also shown. The background is a half-transparent topographic map of Geospatial Information Authority of Japan (GSI) that is underlain by a slope map produced from 5-m grid digital elevation models (DEMs) of GSI

Fig. 14.2 Photograph of southeast-side-down deformation on a paved road at Loc. 423b. Looking to the north. Red arrows indicate the fault trace

Fig. 14.3 Photograph of southeast-side-down deformation of the ground surface immediately southwest of Loc. 423b. Looking to the north. Red arrows indicate the fault trace

Fig. 14.4 Photograph of southeast-side-down deformation of rice paddy at Loc. 423a. Looking to the north. Red arrows indicate the fault trace

Fig. 14.5 Photograph of right-lateral offset (amount 1.5 cm) with northwest-side-up vertical displacement (amount 5 cm) of a paved road at Loc. 423. Looking to the west. Red arrows indicate the fault trace. White arrow indicates a pit excavation site by Ishimura et al. (2017)

Fig. 14.6 Photograph of southwest-side-down deformation and open cracks on a paved road at Loc. 422b. Looking to the north

Fig. 14.7 Photograph of open cracks on a paved road at Loc. 422a. Looking to the west

Fig. 14.8 Photograph of right-lateral offset (amount 5 cm) of block wall at Loc. 422. Looking to the east. Red arrows indicate the fault trace

Fig. 14.9 Photograph of northwest-side-down deformation on a paved road at Loc. 421. Looking to the east. Red arrows indicate the fault trace. A pool of water indicates the downthrown side

Fig. 14.10 Photograph of collapsed tombstones and open cracks at Loc. 421. Looking to the north. Red arrows indicate the fault trace

Fig. 14.11 Photograph of northwest-side-down deformation (amount 10 cm) on a paved parking area at Loc. 420. Looking to the south. Red arrows indicate the fault trace

Fig. 14.12 Photograph of open cracks on a paved road at Loc. 421e. Looking to the south

Fig. 14.13 Photograph of southeast-side-down deformation on a paved parking area at Loc. 421d. Looking to the west. Red arrows indicate the fault trace. A pool of water indicates the downthrown side

Fig. 14.14 Photograph of northwest-side-down offset of greenhouses at Loc. 421c. Looking to the west

Fig. 14.15 Photograph of open cracks on a paved road at Loc. 421b. Looking to the south

Fig. 14.16 Photograph of right-lateral offset (amount 2 cm) of the sideline and open cracks on a paved road at Loc. 421a. Looking to the west

zone. At Loc. 421d, southeast-side-down vertical displacement was identified on a paved parking area (Fig. 14.13). On the other hand, northwest-side-down vertical displacement was identified using the height of the frame of the greenhouses at Loc. 421c (Fig. 14.14).

Recent studies (Fukushima and Ishimura 2020; Ishimura et al. 2021) investigated these surface ruptures using InSAR and paleo-seismic trenching. Fukushima and Ishimura (2020) created 3D deformation maps using InSAR and GNSS data and conducted the inversion modeling of InSAR analysis to estimate the dip angles of both strands and slip distributions on their fault planes. Ishimura et al. (2021) excavated a trench on the northeast part of the north strand (Loc. 423a) and dated the penultimate faulting event at ~2 ka, which is consistent with that of the primary Futagawa fault.

References

Fujiwara S, Yarai H, Kobayashi T, Morishita Y, Nakano T, Miyahara B, Nakai H, Miura Y, Ueshiba H, Kakiage Y, Une H (2016) Small-displacement linear surface ruptures of the 2016 Kumamoto earthquake sequence detected by ALOS-2 SAR interferometry. Earth Planets Space 68:160. https://doi.org/10.1186/s40623-016-0534-x

Fukushima Y, Ishimura D (2020) Characteristics of secondary-ruptured faults in the Aso Caldera triggered by the 2016 Mw 7.0 Kumamoto earthquake. Earth Planets Space 72:175. https://doi.org/10.1186/s40623-020-01306-y

Ishimura D, Toda S, Ichihara T, Takahashi N, Konno A, Sato H (2017) A study on surface ruptures around Miyaji, Aso City, Kumamoto Prefecture, associated with the 2016 Kumamoto earthquake sequence and upward slip tapering on pit excavation walls. Active Fault Res 47:9–16. https://doi.org/10.11462/afr.2017.47_9

Ishimura D, Tsutsumi H, Toda S, Fukushima Y, Kumahara Y, Takahashi N, Ichihara T, Takada K (2021) Repeated triggered ruptures on a distributed secondary fault system: an example from the 2016 Kumamoto earthquake, southwest Japan. Earth Planets Space 73:39. https://doi.org/10.1186/s40623-021-01371-x

Surface Ruptures in the Northwest of the Outer Aso Caldera

15

Hiroshi Une, Takayuki Nakano, Satoshi Fujiwara, Hiroshi P. Sato, and Hiroshi Yagi

Abstract

Fujiwara et al. (2016) detected the surface displacement field associated with the 2016 Kumamoto earthquake sequence from the SAR interferometric analysis using ALOS-2 SAR images and mapped linear discontinuities in the interferograms as the "displacement lineaments (DLs)." Dozens of parallel DLs were pointed out in the area northwest of the outer rim of the Aso Caldera, running from west–northwest to east–southeast. The length of each DL is predominantly 3–5 km. The horizontal spacing between each rupture is approximately 0.5–1 km, and the displacements judged from the discontinuity in the interferograms are primarily around 10-cm dip slip. While many of the DLs appeared along the known active faults, some of them appeared where there are no tectonic geomorphic features. We conducted a series of field surveys to clarify the surface deformation of these DLs in situ and found more than ten surface ruptures along DLs on the ground. The displacement of these ruptures was less than a few tens centimeters, concordant with observations of Fujiwara et al. (2016) in terms of positions, directions, and slip senses. In this chapter, nine examples of such ruptures are described in detail with the photos taken on the ground.

Keywords

SAR interferometry • ALOS-2 • Displacement lineament • Outer rim of the Aso Caldera • In situ field survey

Supplementary Information
The online version contains supplementary material available at (https://doi.org/10.1007/978-981-19-1150-7_15).
Note
The supplementary material of this book contains geographical information system (GIS) and spreadsheet files of the surface ruptures associated with the mainshock of the 2016 Kumamoto earthquake sequence

H. Une (✉)
AI Data Science Center, Ochanomizu University, 2-1-1, Otsuka, Bunkyo-ku, Tokyo, 112-8610, Japan
e-mail: hiroune@nifty.com

T. Nakano
Geographic Department, Geospatial Information Authority of Japan, 1, Kitasato, Tsukuba, Ibaraki, 305-0811, Japan
e-mail: nakano-t96fj@mlit.go.jp

S. Fujiwara
Geodetic Department, Geospatial Information Authority of Japan, 1, Kitasato, Tsukuba, Ibaraki, 305-0811, Japan
e-mail: fujiwara-s2vq@mlit.go.jp

H. P. Sato
Department of Geography, College of Humanities and Sciences, Nihon University, 3-25-40, Sakurajosui, Setagaya-ku, Tokyo, 156-8550, Japan
e-mail: satou.hiroshi37@nihon-u.ac.jp

H. Yagi
Faculty of Education, Art and Science, Yagamagata University, 1-4-12, Kojirakawa, Yamagata, Yamagata, 990-8560, Japan
e-mail: yagi@e.yamagata-u.ac.jp

Numerous scattered small surface ruptures associated with the Kumamoto earthquake sequence were found in the area northwest of the outer rim of the Aso Caldera, 5–20 km away from the Futagawa fault. It was difficult to illustrate the overall distribution of these ruptures by field investigation or aerial photo interpretation because they appeared in the mountainous area, and the displacement of ruptures was too small, less than a few tens centimeters.

Fujiwara et al. (2016) mapped out the "linear surface ruptures" detected from SAR interferometric (InSAR) analysis. They detected the displacement field on the ground surface associated with the 2016 Kumamoto earthquake sequence using ALOS-2 SAR images. We use the word "displacement lineament (DL)" for "linear surface rupture" hereinafter to avoid the confusion with surface ruptures on the ground identified by in situ surveys (Fujiwara et al. 2020). The DLs were derived from linear phase discontinuities and/or offsets, showing displacement in the

Y. Kumahara et al. (eds.), *Surface Ruptures Associated with the 2016 Kumamoto Earthquake Sequence in Southwest Japan*, Advances in Geological Science, https://doi.org/10.1007/978-981-19-1150-7_15

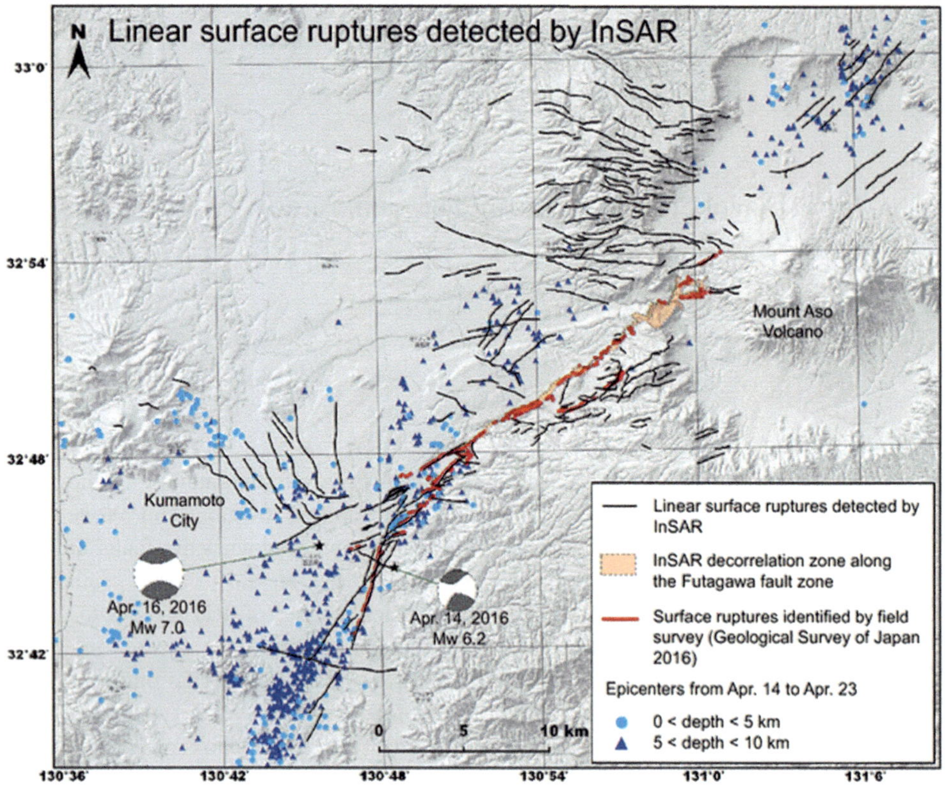

Fig. 15.1 Displacement lineaments identified from InSAR analysis (Fujiwara et al. 2016)

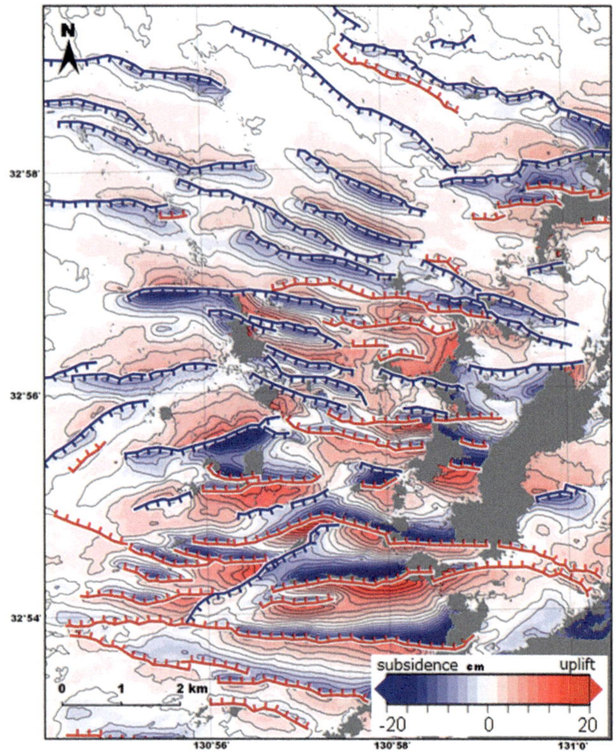

Fig. 15.2 High-pass-filtered up-down displacement map. Three-dimensional (3D) analysis was after Morishita et al. (2016). Red and blue areas represent up and down, respectively. The contour interval is 2 cm. Red solid lines show the identified displacement lineaments with dip slip where the north side moved down, and solid blue lines show the identified displacement lineaments with dip slip where the south side moved down. Ticks of the lines show the lower side of each displacement (Fujiwara et al. 2016)

interferograms. Fujiwara et al. (2016) pointed out dozens of parallel DLs in the area northwest of the outer rim of the Aso Caldera, running from west–northwest to east–southeast (Fig. 15.1). The length of each DL is predominantly 3–5 km. The horizontal spacing between each rupture is approximately 0.5–1 km, and the displacements judged from the discontinuity in the interferograms are primarily around 10-cm dip slip (Fig. 15.2). While many of the DLs appeared along the known active faults, some of them appeared where there are no tectonic geomorphic features. Fujiwara et al. (2016) inferred that many of the DLs were caused by normal faulting of half grabens induced by north–south extension associated with the earthquake.

We conducted a series of field surveys to clarify the surface deformation of these DLs in situ and found more than ten surface ruptures along DLs on the ground. Most of such surface ruptures are concordant with observations of Fujiwara et al. (2016) in terms of positions, directions, and slip senses (Fig. 15.3). Some examples of such ruptures are described as follows.

(1) **Former Yagoyama Golf Park** (Loc. 1)

A surface rupture continued from the parking lot in front of the clubhouse to the fairway of the Yagoyama Golf Park (Fig. 15.4), which was operated at the time of the Kumamoto earthquake but abandoned in June 2017 and later transferred to a solar power plant. The length of the surface rupture was at least 250 m, and the strike was approximately east–west. The vertical displacement was all north-side down, and the maximum amount of displacement was 18 cm at the parking lot (Fig. 15.5) and a few cm at the fairway (Fig. 15.6). The rupture ran across a pond in the fairway, and

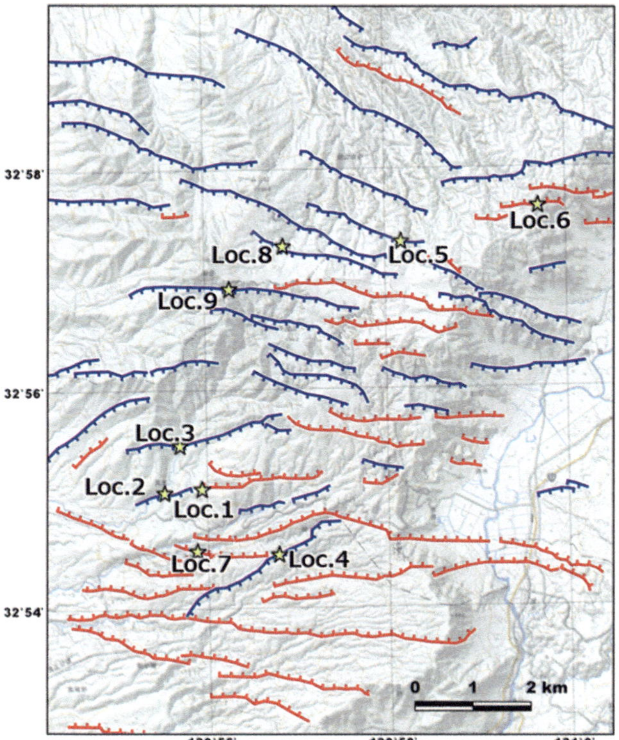

Fig. 15.3 Displacement lineaments (Fujiwara et al. 2016) and locations of major surface ruptures identified on the ground. Red solid lines show the displacement lineaments with dip slip where the north side moved down, and solid blue lines show the identified displacement lineaments with dip slip where the south side moved down. Ticks of the lines show the lower side of each displacement. Yellow stars indicate locations where major or characteristic surface ruptures were identified on the ground

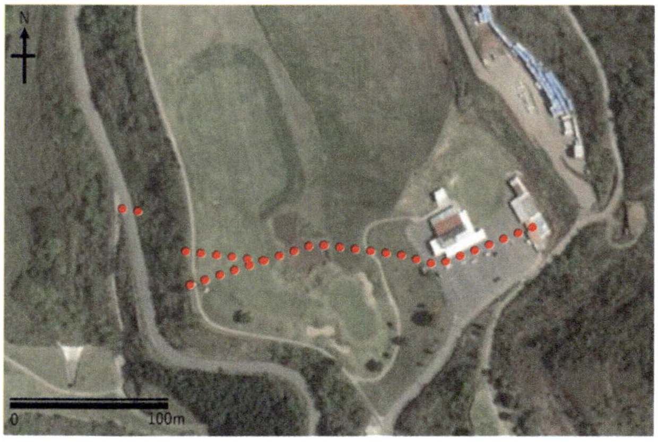

Fig. 15.4 Surface rupture traces at the former Yagoyama Golf Park (Loc. 1) identified by field survey (red-dotted lines) plotted over an orthophotograph taken on April 20, 2016 by the Geospatial Information Authority of Japan

Fig. 15.5 Photograph of the surface rupture at the parking lot of the former Yagoyama Golf Park (Loc. 1). Looking from west to east

Fig. 15.6 Photograph of the surface rupture at the fairway of the former Yagoyama Golf Park (Loc. 1). Looking from east to west

Fig. 15.7 Photograph of the surface rupture that cuts across the riverbed sill and sidewalls of the Yago River (Loc. 2). Yellow arrows indicate the location of the rupture

the water leaked out to make it dry. The rupture branches into two traces in the fairway. A north-side-drop crack was observed on an asphalt road immediately west of the golf course (Fig. 15.4). If it is the continuation of the same rupture, it suggests that the rupture was formed by normal faulting on a north-dipping fault plane, which is consistent with the displacement inferred from InSAR analysis. The position of this rupture is on a geomorphologically evident active fault, a branch fault of the Sakanoshita fault (Research Group for Active Tectonics in Kyushu 1989), shown in Suzuki et al. (2017).

(2) **Maki, Ozu Town** (Loc. 2)

A clear surface rupture appeared across multiple artificial features including a small concrete paved road, the riverbed sill and sidewalls of the Yago River (Fig. 15.7), an asphalt road (Fig. 15.8), underfloor of a house, and another asphalt

Fig. 15.8 Photograph of the surface rupture on an asphalt road at Maki, Ozu Town (Loc. 2). Photo by Soba Cafe Yago-an

Fig. 15.9 Photograph of the surface rupture near the gate of former Hinohara Campsite (Loc. 3)

Fig. 15.10 Photograph of the surface rupture on the slope of the cedar forest at Toge, Ozu Town (Loc. 4)

road from east to west. The length of the rupture was at least 100 m. The strike was approximately east–west, and the vertical displacement was 10 cm at maximum at the asphalt road, down to the south. It is located on the western extension of the DL on which the surface rupture at Loc. 1 was found. However, the sense of the vertical displacement was opposite, which coincides with the result of the InSAR analysis (Figs. 15.1 and 15.2).

(3) **Near the Gate of Former Hinohara Campsite** (Loc. 3)

A 20-cm south-side-drop vertical displacement was found across a gravel road near the abandoned Hinohara campsite entrance (Fig. 15.9). The strike was approximately east–west. It is situated just on one of the DLs shown in Fujiwara et al. (2016).

Fig. 15.11 Photograph of the surface rupture in the forest west of Matoishi-Hatabe Bokuya (Loc. 5)

Fig. 15.12 Photograph of the surface rupture in the grass field beside the Milk Road at Adogase Bokuya, Aso City (Loc. 6)

Fig. 15.13 Photograph of the surface rupture on an asphalt road at Furujo, Ozu Town (Loc. 7). It possibly included a 10-cm left-lateral offset

(4) **Toge, Ozu Town** (Loc. 4)

On the slope of the cedar forest, an east–west-trending open crack continued for at least several tens of meters. The width of the opening is 12 to 40 cm, and the vertical displacement with a south drop of 10 to 20 cm was observed (Fig. 15.10). The scarp faced uphill, indicating that the gravitational movement cannot explain the rupture.

(5) **West of Matoishi-Hatabe Bokuya** (Loc. 5)

An east–west-trending surface rupture with a south drop vertical displacement of about 25 cm was found across a forest road. In the forest on the west extension of this rupture, surface ruptures with displacements of about 40 cm at maximum continued intermittently for more than 100 m further to the west (Fig. 15.11). The position of this rupture is on the known active fault, the Hatabe-genya fault (Research Group for Active Tectonics in Kyushu 1989) or Matoishi-Hatabe III fault shown in Suzuki et al. (2017).

(6) **The Milk Road at Adogase Bokuya, Aso City** (Loc. 6)

We identified the trace of pavement repair to smooth out the north-side-down sharp drop of the road surface caused by the earthquake. The ditch beside the road also showed vertical displacement of about 8 cm. In the grass field on the west extension of this trace, we observed a clear north-side drop surface rupture trending approximately east–west with a maximum vertical displacement of about 50 cm (Fig. 15.12). It continued for at least 100 m.

(7) **Furujo, Ozu Town** (Loc. 7)

A surface rupture with a north-side drop displacement of about 15 cm was identified on the asphalt road. The strike was approximately east–west. It possibly included a 10-cm left-lateral offset (Fig. 15.13). The position of this rupture is on the known active fault, the Furujo fault (Research Group for Active Tectonics in Kyushu 1989), also shown in Suzuki et al. (2017).

(8) **East of Kuradake** (Loc. 8)

At the foot of the south-facing fault scarp of the Matoishi-bokujo I fault (Research Group for Active Tectonics in Kyushu 1989; Suzuki et al. 2017), a few cm opening of the surface soil was identified although clear surface rupture was not found. Fujiwara et al. (2016) showed that the DL along the active fault has a north-side-up displacement, which coincides with the sense of the cumulative displacement of the Aso-4 pyroclastic flow surface due to the faulting. At this site, a trench excavation survey and ground-penetrating radar (GPR) survey were executed at the foot of the fault scarp (Nakano et al. 2018; Une et al. 2019; Sato et al. 2021). These surveys suggested that the displacement took place cumulatively to form the fault topography. However, it is unknown whether such cumulative displacement is due to seismogenic faulting or passive activity similar to that during the 2016 earthquake.

(9) **Southeast of Kuradake** (Loc. 9)

An N80°W-trending surface rupture with south-side drop vertical displacement of about 25 cm was found on the slope beside a road. On the eastern and western extensions of this rupture, the traces of the pavement repair were identified on the road, suggesting about 45-cm-vertical displacement.

Acknowledgements A part of this work was supported by JSPS KAKENHI Grant Number JP17K01234 and 20K01141.

References

Fujiwara S, Nakano T, Morishita Y (2020) Detection of triggered shallow slips caused by large earthquakes using L-band SAR interferometry. Earth Planets Space 72:119

Fujiwawa S, Yarai H, Kobayashi T, Morishita Y, Nakano T, Miyahara B, Nakai H, Miura Y, Ueshiba H, Kakiage Y, Une H (2016) Small-displacement linear surface ruptures of 2016 Kumamoto earthquake sequence detected by ALOS-2 SAR interferometry. Earth Planets Space 68:160

Morishita Y, Kobayashi T, Yarai H (2016) Three-dimensional deformation mapping of a dike intrusion event in Sakurajima in 2015 by exploiting the right- and left-looking ALOS-2 InSAR. Geophys Res Lett 43:4197–4204

Nakano T, Une H, Sato HP, Komura K, Yagi H (2018) Geomorphological and geological characteristics of triggered surface faults associated with 2016 Kumamoto Earthquake sequence around the northwest of the outer rim of the Aso caldera, southwestern Japan. In: Abstracts of AGU Fall Meeting 2018, Walter E Washington Convention Center, Washington, DC, 10–14 Dec 2018

Research Group for Active Tectonics in Kyushu (ed) (1989) Active Tectonics in Kyushu. University of Tokyo Press, Tokyo

Sato HP, Komura K, Une H, Nakano T, Yagi H (2021) Study on cumulative activities of passively ruptured faults through a trenching survey at the Matoishi Bokujo I Fault, northwest side of the Aso caldera, southwestern Japan. Geogr Rev Jpn Ser A 94(4):250–264

Suzuki Y, Ishimura D, Kumaki Y, Kumahara Y, Chida N, Nakata T, Nakano T (2017) 1:25,000 active fault map, Aso. Geospatial Information Authority of Japan

Une H, Nakano T, Sato HP, Yagi H, Komura K (2019) Past activity of 'triggered fault' in the northwest of outer rim of Aso caldera estimated from the results of excavation survey and displacement topography. In: Abstracts of the Japan Geoscience Union Meeting 2019, Makuhari Messe, Chiba, 26–30 May 2019

Surface Ruptures Along the Western Part of the Bungo Kaido Road

16

Mitsuhisa Watanabe, Takashi Nakata, Yasuhiro Suzuki, Yasuhiro Kumahara, and Kei Tanaka

Abstract

The surface ruptures with consistent right-lateral slip components appeared in the ENE-WSW direction intermittently along the western part of the Bungo Kaido road during the 2016 Kumamoto earthquake. The ruptures seem to have appeared in severely damaged zones, indicating that a strong ground motion was generated in a very shallow part.

Keywords

Bungo Kaido road • Kumamoto earthquake • Surface rupture • Right-lateral slip • Strong ground motion

The Bungo Kaido road crossing central Kyushu was an important road for "Sankin-kotai" (alternate attendance by feudal lords such as Kiyomasa Kato) and economic activities in the Edo period. The "Seishokodo road (Kiyomasa Kodo road)" constituted the western part of the Bungo Kaido road (Fig. 16.1). The Seishokodo road was constructed by digging the plateau surfaces and is currently being reclaimed. During the 2016 Kumamoto earthquake, surface ruptures appeared along the western part of the Bungo Kaido road, as described below.

We confirmed the surface ruptures at a crossroad close to the western caldera rim of the Aso volcano, where the Bungo Kaido road and the prefectural route 23 meet (Loc. B1). The prefectural route was broken, and the wall of a drainage channel along the route was right-laterally offset 2 cm (Fig. 16.2). The wall had already been repaired.

Around the stele of the Seishokodo road, a surface rupture appeared on a stone pavement (Loc. B2). The rupture showed several cm north-up displacement with no lateral offset (Fig. 16.3). The stele fixed with four bolts fell out of the pedestal and toppled northward (Fig. 16.4). West of Loc. B2, the amount of north-up vertical displacement was 10 cm, and the other stone monuments and small arbor also collapsed in the same direction (Loc. B3). Some buried stone pillars popped out near the rupture (Fig. 16.5). These observations indicate that strong ground motions occurred in the area.

At Shinkoya (Loc. B4; Fig. 16.6), the vertical separation of the surface rupture was several cm to 10 cm up on the south. The foundation of a meeting house was right-laterally offset by 5 cm (Fig. 16.7). In addition, the rupture displayed a left-stepping en-echelon arrangement of cracks (Loc. B5;

Supplementary Information
The online version contains supplementary material available at (https://doi.org/10.1007/978-981-19-1150-7_16).

Note
The supplementary material of this book contains geographical information system (GIS) and spreadsheet files of the surface ruptures associated with the mainshock of the 2016 Kumamoto earthquake sequence

M. Watanabe (✉)
Faculty of Sociology, Toyo University, 5-28-20, Hakusan, Bunkyo-ku, Tokyo, 112-8606, Japan
e-mail: manqq@toyo.jp

T. Nakata
Department of Geography, Hiroshima University, 1-2-3, Kagamiyama, Higashi-Hiroshima, Hiroshima, 739-8522, Japan

Y. Suzuki
Disaster Mitigation Research Center, Nagoya University, Furo-cho, Chikusa-ku, Nagoya, Aichi, 464-8601, Japan
e-mail: ysz@nagoya-u.jp

Y. Kumahara
Graduate School of Humanities and Social Sciences, Hiroshima University, 1-1-1, Kagamiyama, Higashi-Hiroshima, Hiroshima, 739-8524, Japan
e-mail: kumakuma@hiroshima-u.ac.jp

K. Tanaka
College of Economics, Nihon University, 1-3-2, Kanda-Misakicho, Chiyoda-ku, Tokyo, 101-8360, Japan
e-mail: tanaka.kei@nihon-u.ac.jp

Y. Kumahara et al. (eds.), *Surface Ruptures Associated with the 2016 Kumamoto Earthquake Sequence in Southwest Japan*, Advances in Geological Science, https://doi.org/10.1007/978-981-19-1150-7_16

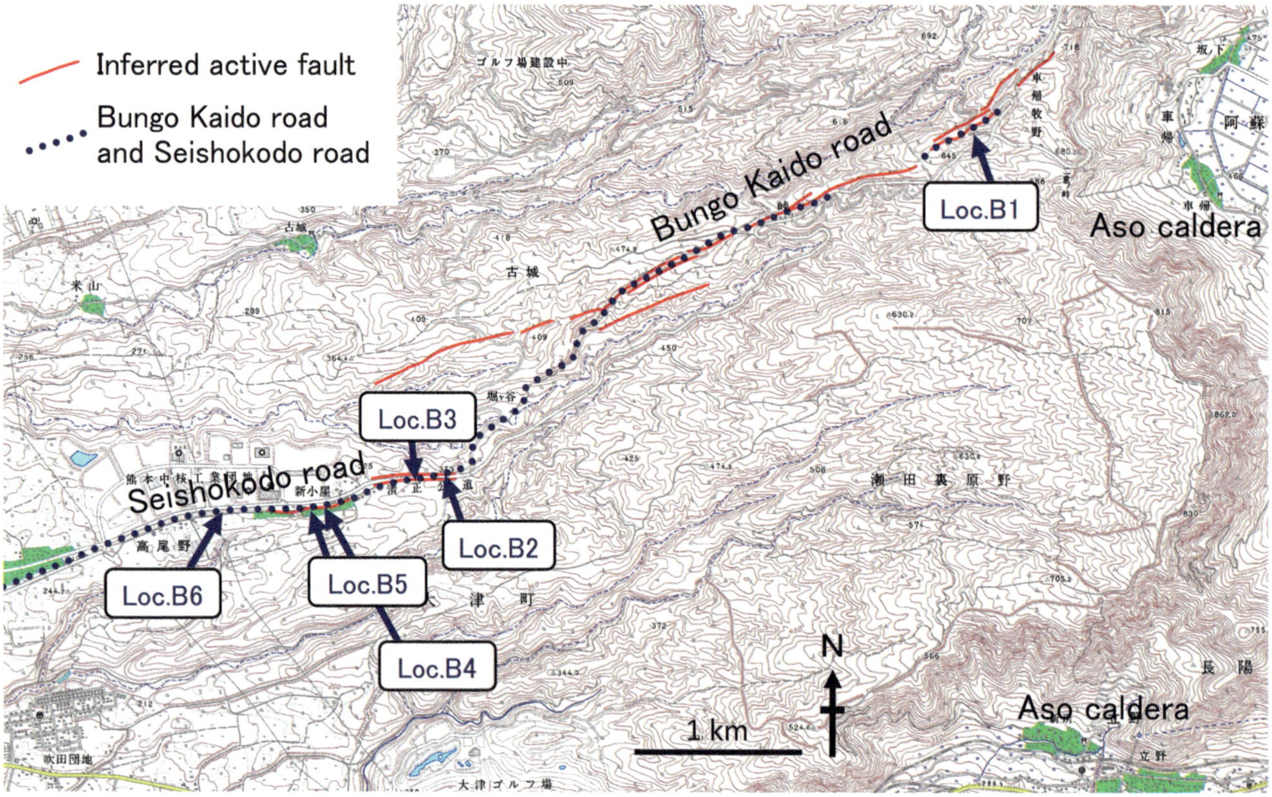

Fig. 16.1 Surface rupture map along the Bungo Kaido road. The background is a 1:25,000 scale topographic map of the Geospatial Information Authority of Japan

Fig. 16.2 Photograph of the right-lateral offset of drainage channel wall at Loc. B1. The thick-red arrows indicate the location of the surface break

Fig. 16.3 Photograph of the surface rupture on the stone pavement at Loc. B2. Red arrows indicate the trace of the rupture. The rupture had several cm north-up vertical separation

Fig. 16.4 Photograph of the toppled stele of the Seishokodo road at Loc. B2. Strong ground motion defeated the stele that had been fixed with four bolts

Fig. 16.5 Photograph of the stone pillars that jumped out of the ground at Loc. B3. The red arrow indicates the location of the surface break. The rupture had north-up vertical separation with an amount of ~10 cm. The strong ground motion pulled out the stone pillars and destroyed the stone monument and small arbor

Fig. 16.8), indicating the right-lateral fault movement. The rupture was recognized by ~10 cm south-up vertical separation. Some wide open cracks appeared, and a light truck wheel had fallen into one of them. The surface break could be traced for at least 120 m, and damage to the houses along the trace was notable.

At Loc. B5, we emplaced a trench across the en-echelon cracks (Fig. 16.9). The deposits exposed on the trench wall were classified into five units (A–E). Unit A is the artificial fill of the Seishokodo road. Units B and D are organic black soils. The radiocarbon sample collected from the lower part of unit D was dated at 21,940 ± 70 yBP (Beta-517835).

Fig. 16.6 Photograph of the surface rupture at Loc. B4. The red arrow indicates the trace of the surface fault. The ground failure damaged a meeting house. The south side of the trace was raised about 10 cm

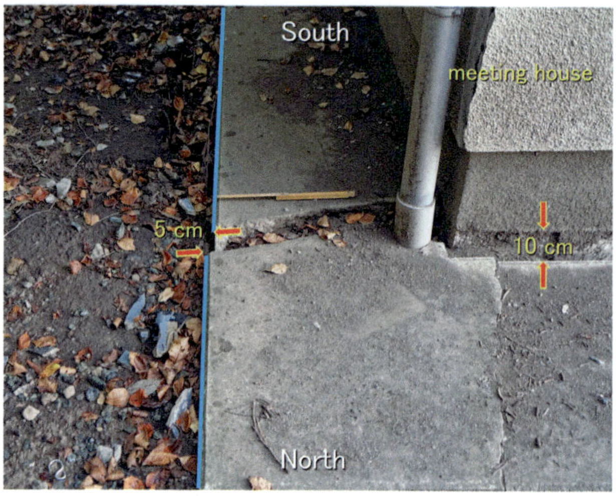

Fig. 16.7 Photograph of the right-lateral offset of the foundation of the meeting house at Loc. B4

Unit C is assumed to be the K-Ah tephra (ca. 7.3 ka; Machida and Arai 2003). Unit E is a brown aeolian deposit intercalated with scoria and pumice layers erupted from the central volcanoes within the Aso caldera. The surface crack is located close to the southern edge of the artificial fill materials (unit A). If unit A moved by shaking during the 2016 Kumamoto earthquake, it should have moved left-laterally based on the height of the ground surface. It is entirely unreasonable that the fill moved right-laterally. There were also many fissures in unit B to unit E layers just below the surface crack. These fissures may have been caused by right-lateral faulting.

To the west of Shinkoya (Loc. B6), surface ruptures appeared intermittently for several tens of meters. A small-scale pull-apart basin formed by dextral strike-slip faults extending ESE-WNW direction (Fig. 16.10). The width of the pull-apart basin was 1.6 m, and the length was over 3.5 m. The vertical displacement of the rupture bounding the basin was ~8 cm at the maximum.

The surface ruptures with consistent right-lateral slip component appeared in ENE-WSW direction intermittently along the Bungo Kaido road. In addition, we observed north-up vertical separation in the eastern part and south-up vertical separation in the western part, respectively.

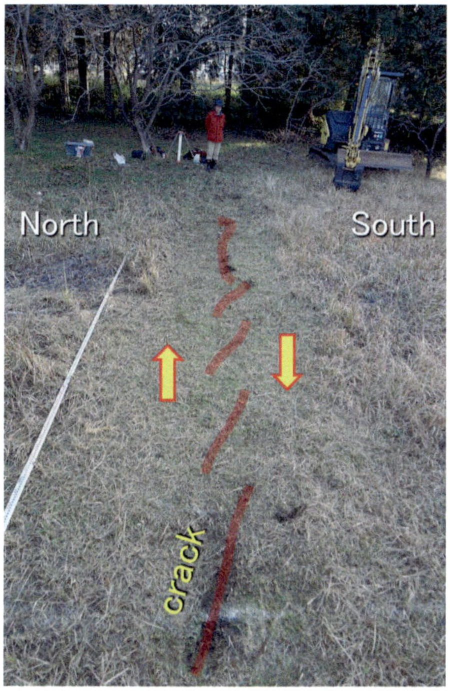

Fig. 16.8 Photograph of the left-stepping en-echelon arrangement of cracks at Loc. B5. The amount of lateral displacement was unknown. The rupture had south-up vertical separation with an amount of ~10 cm

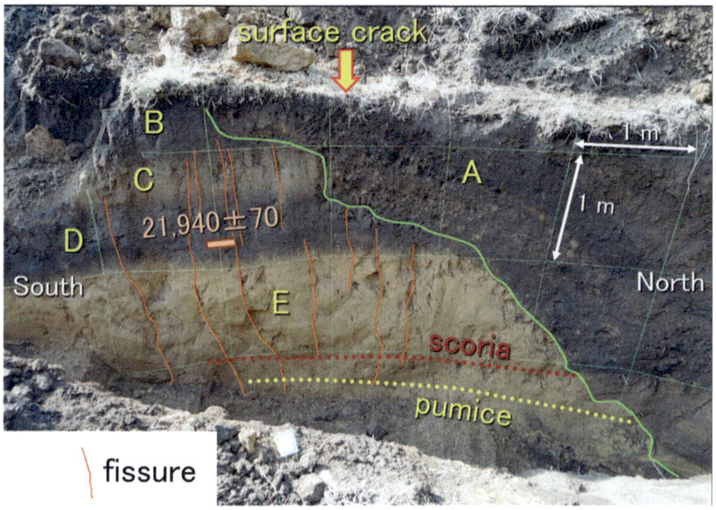

Fig. 16.9 Photograph and log of the west wall of the trench at Loc. B5. Unit A is the fill of the Seishokodo road. Units B and D are organic black soils. Units C and E are volcanic ash and brown aeolian deposit, respectively. The surface crack is located close to the boundary between units A and B, and there are many fissures in unit B to unit D just below the surface crack

It coincides with the pattern of vertical separation along a single right-lateral strike-slip fault. However, we cannot preclude the possibility that the surface cracks were caused by the lateral flow of the artificial fill of the Seishokodo road.

A narrow, severely damaged zone seems to have appeared along the Bungo Kaido road during the 2016 Kumamoto earthquake. If so, the strong ground motion should have been generated in a very shallow part. The depth to the top of the seismogenic layer needs to be carefully reviewed.

Acknowledgements The authors wish to thank Tsunayoshi Funai for permission to access his land. We also thank Kenji Takahama for his kind advice during excavation.

Fig. 16.10 Photograph of the small-scale pull-apart basin formed by right-lateral slip at Loc. B6. Red arrows indicate the traces of surface faults. The maximum amount of vertical displacement was ~8 cm

Reference

Machida H, Arai F (2003) Atlas of tephra in and around Japan [revised edition], University of Tokyo Press, Tokyo

Surface Ruptures in Mashiki Town: Tectonic Significance and Building Damage

17

Yasuhiro Suzuki, Mitsuhisa Watanabe, and Takashi Nataka

Abstract

During the 2016 Kumamoto earthquake, surface fault ruptures appeared not only along the well-known active Futagawa-Hinagu fault but also in the urbanized area of Mashiki Town, where no active faults have previously been identified. This chapter describes the distribution of the earthquake faults in Mashiki Town and evaluates their relationship with tectonic landforms and building damage based on Suzuki et al. (2018). We conclude that: (1) The earthquake faults were continuous for at least 5 km from the main trace of the Futagawa fault; (2) The fault diverged from east to west but maintains the total amount of displacement of approximately 70 cm; (3) The earthquake fault locations coincided with those of tectonic landforms, and (4) building damage tended to be severe near the earthquake fault.

Supplementary Information
The online version contains supplementary material available at (https://doi.org/10.1007/978-981-19-1150-7_17).
Note
The supplementary material of this book contains geographical information system (GIS) and spreadsheet files of the surface ruptures associated with the mainshock of the 2016 Kumamoto earthquake sequence

Y. Suzuki (✉)
Disaster Mitigation Research Center, Nagoya University, Furo-cho, Chikusa-ku, Nagoya, Aichi, 464-8601, Japan
e-mail: ysz@nagoya-u.jp

M. Watanabe
Faculty of Sociology, Toyo University, 5-28-20, Hakusan, Bunkyo-ku, Tokyo, 112-8606, Japan
e-mail: manqq@toyo.jp

T. Nataka
Department of Geography, Hiroshima University, 1-2-3, Kagamiyama, Higashi-Hiroshima, Hiroshima, 739-8522, Japan
e-mail: tnakata@hiroshima-u.ac.jp

Keywords

2016 Kumamoto earthquake • Earthquake fault • Active fault • Building damage

17.1 Emergence of a Previously Unknown Surface Fault

A previously unknown surface fault trace emerged in Mashiki Town during the 2016 Kumamoto earthquake, which diverged from the Futagawa fault in the Dozon area (Fig. 17.1). The trace extended to the urbanized area and caused severe damage to buildings. The fault trace was clearly visible in a 3.5-km section east of Mashiki Town, with cracks and displacements observed in paddy fields. However, the fault trace was hard to observe in Mashiki Town due to many collapsed buildings covering the ground surface. Suzuki et al. (2016, 2018) repeated the field survey one year after the earthquake and were able to observe the entire surface trace of the earthquake fault. Their original data were referenced in Fig. 17.1 by Kumahara et al. (2017).

In Mashiki Town, the Japanese Meteorological Agency seismic intensity of 7 (maximum scale) was measured twice on the 14th and 16th of April 2016, resulting in tremendous building damage. The most severe damage occurred during the mainshock on the 16th of April. An area several hundred meters wide along the prefectural road 28, which extends east–west, was severely damaged, similar to the "extremely damaged zone" reported during the 1995 Kobe earthquake (Watanabe et al. 2016).

Some researchers have suggested that the cause of such severe damage was strong ground motion related not to surface faulting but to the specific ground conditions of the site (e.g., Hiroyuki Goto et al. 2017). In addition, the Ministry of Land, Infrastructure, Transport and Tourism (MLIT)

Y. Kumahara et al. (eds.), *Surface Ruptures Associated with the 2016 Kumamoto Earthquake Sequence in Southwest Japan*, Advances in Geological Science, https://doi.org/10.1007/978-981-19-1150-7_17

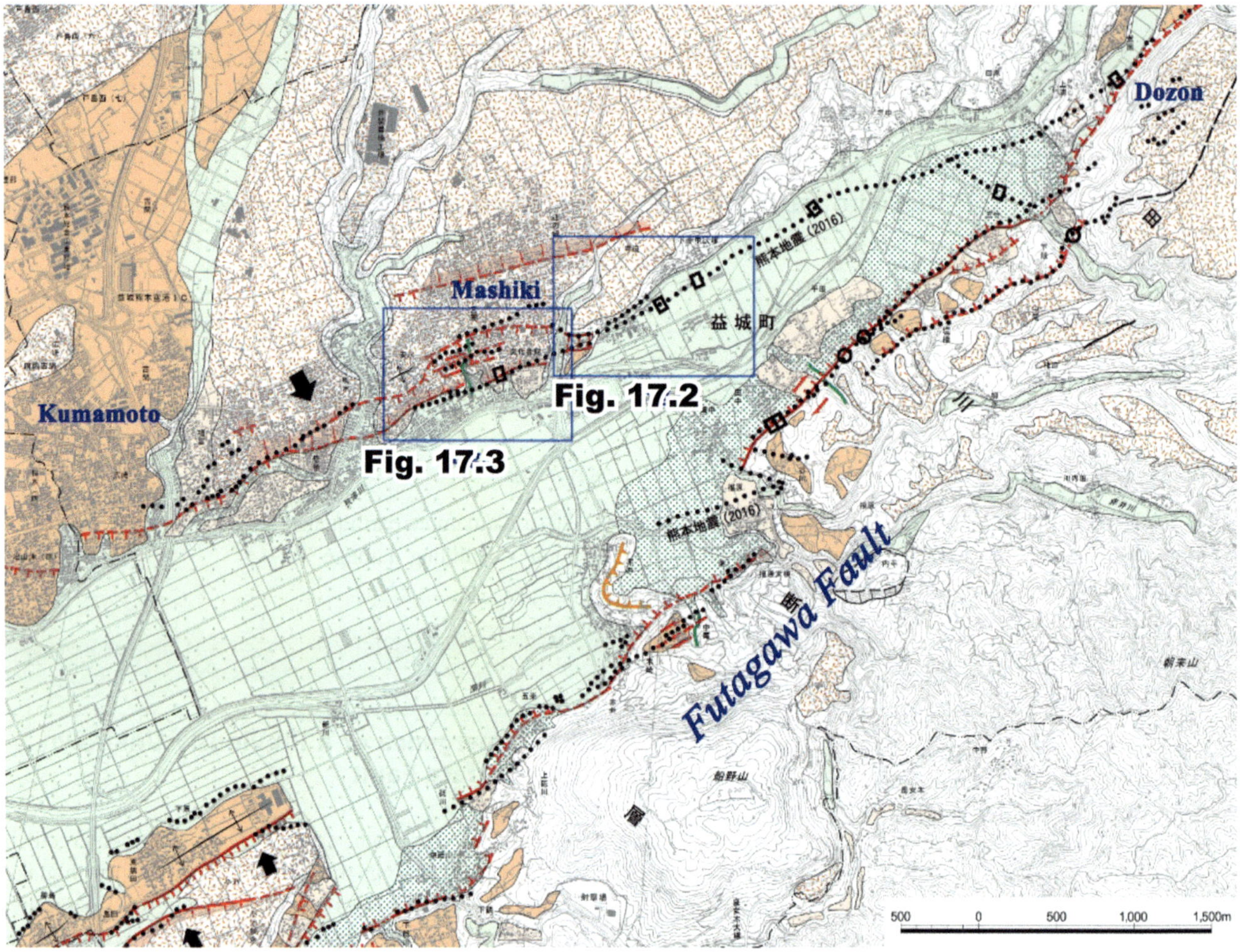

Fig. 17.1 Active fault map "Kumamoto (revised edition)" (Kumahara et al. 2017) and the location of this study

(2017) quickly presented a negative perspective of building damage due to surface faulting. Therefore, this chapter employs a previous report on building damage (Monma et al. 2017) to preliminarily analyze the relationship between surface faulting and building damage in Mashiki Town.

17.2 Description of the Surface Ruptures in Mashiki Town

Earthquake faults were identified according to the principle presented by Nakata et al. (1995) and Suzuki and Watanabe (2006). We focused on ruptures consistent with the seismic fault mechanism, considered the origin of ruptures according to their continuity and regularity and excluded gravitational deformation. We identified earthquake faults in several areas shown in Figs. 17.2 and 17.3 and determined the distribution of fault displacement. The observed earthquake faults are summarized as follows, from east to west.

Right-lateral offsets were confirmed on the alluvial lowland between Locs. T00 and T09 (Fig. 17.2). Left-stepping *en-echelon* faulting was observed between Locs. T03 and T06, which branched into two lines between Locs. T10 and T38. The amount of right-lateral offset was 30 cm and 15–20 cm for the north and south traces, respectively. The surface ruptures on the north were north-side-up, and those on the south were south-side-up, forming a graben with a width of approximately 20–70 m (Figs. 17.4 and 17.5). West of the Terasako street intersection (Loc. T27a), the earthquake fault branched to the west.

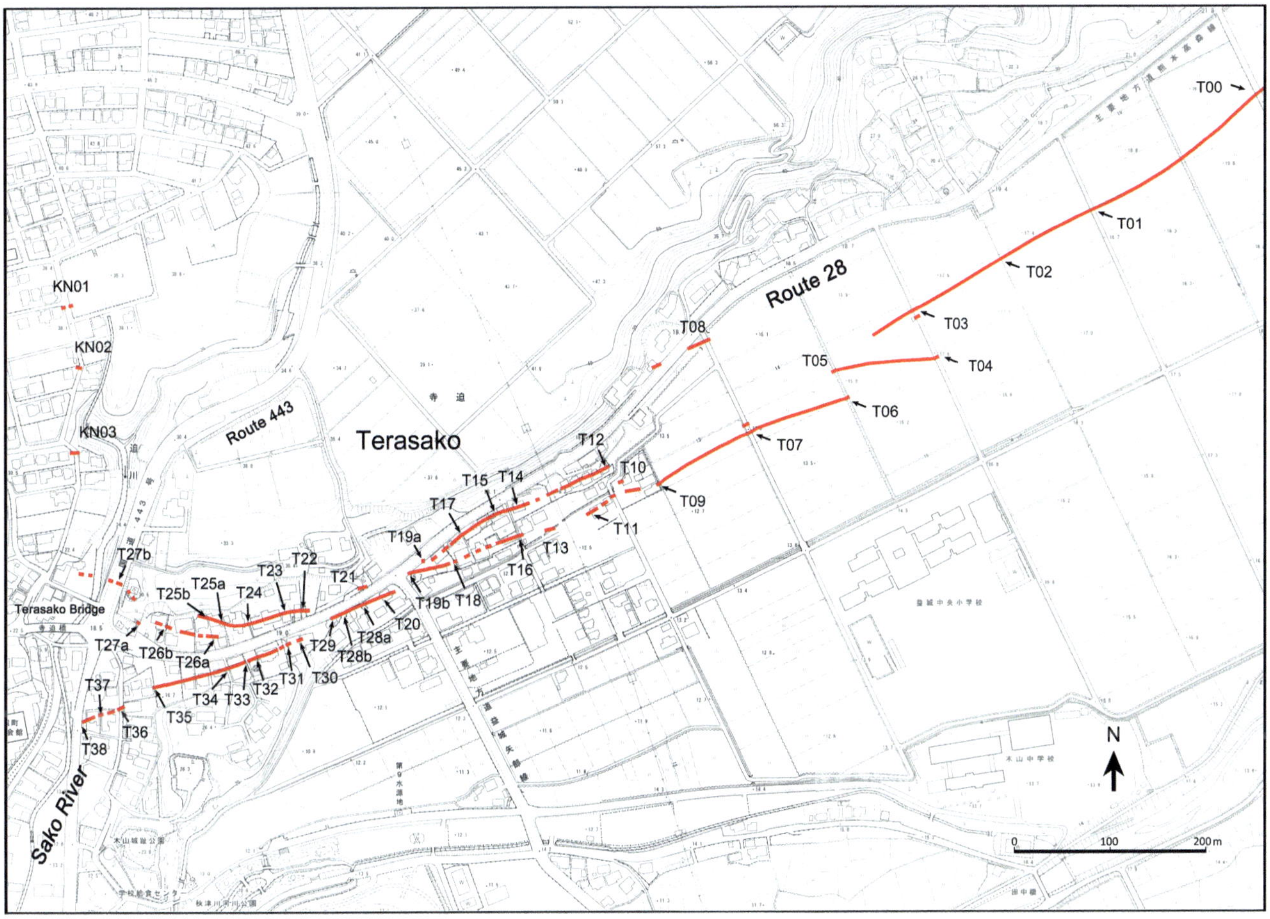

Fig. 17.2 Detailed map of the surface ruptures around Terasako, Mashiki Town (Suzuki et al. 2018). The location of the map is shown in Fig. 17.1

In Kiyama area, there were three traces. The southernmost trace (Locs. KS01 to KS15a, b) had clear continuity with a maximum displacement of 40 cm (Fig. 17.6). The central trace was identified along the prefectural road 28 for 350 m between KC01 and KC09 (Fig. 17.7). As apparent ruptures were unlikely to appear on asphalt-paved roads, some faults may not have been found. The maximum displacement was 30 cm in the right-lateral direction. The northern trace passed through the site of Mashiki Town Hall. The maximum displacement reached 20 cm at KN09 and KN12 (Fig. 17.8).

17.3 Analysis of the Surface Rupture Geometry and Displacement

The earthquake fault was a single right-lateral fault in the paddy fields east of Mashiki Town; however, it branched into two traces in the vicinity of Terasako and further branched into three traces in Kiyama to Miyazono (Figs. 17.2 and 17.3). This divergence (Nakata et al. 1998) strongly suggests that the fault rupture propagated from east to west. This inference seems to contradict the fact that the

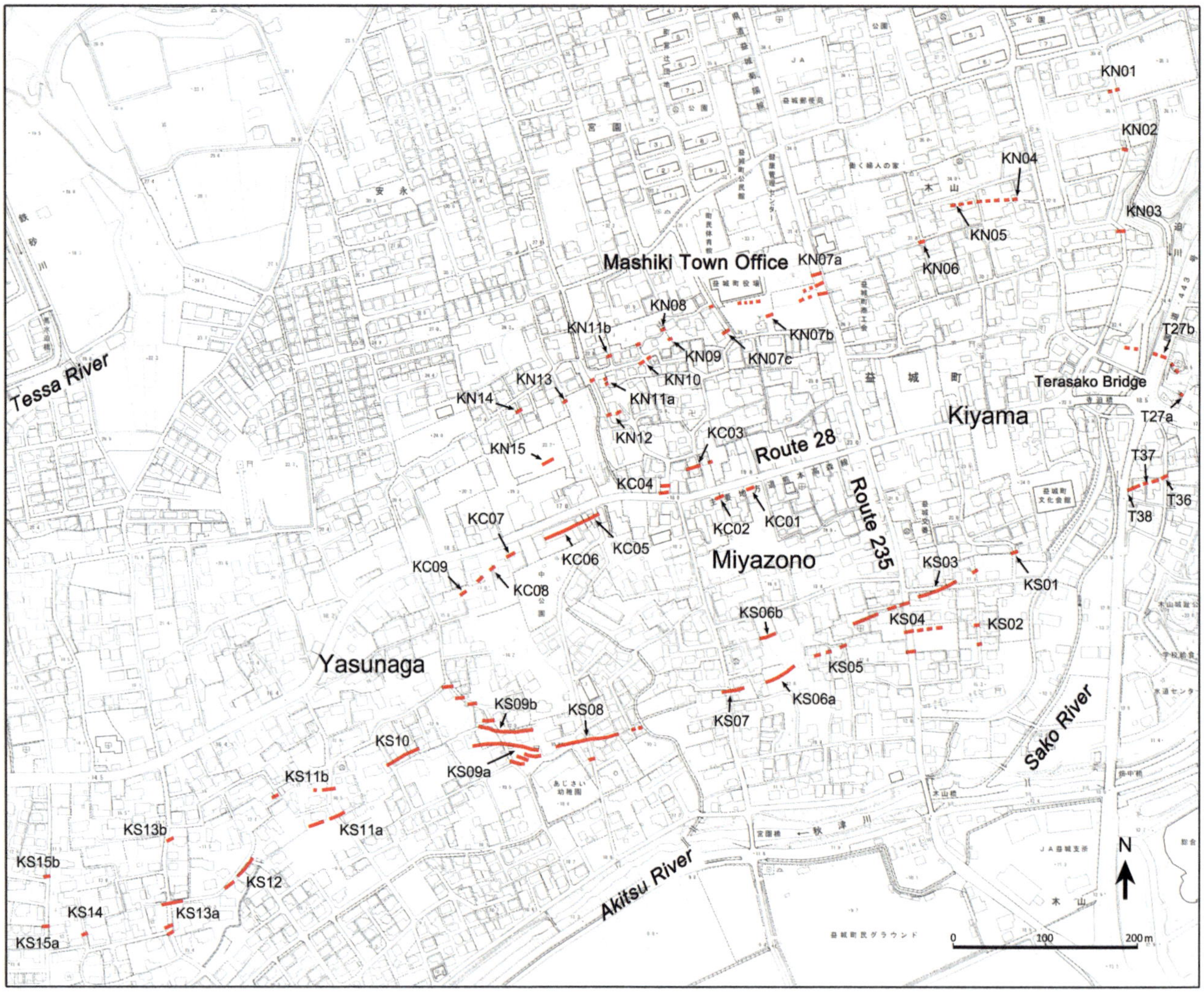

Fig. 17.3 Detailed map of the surface ruptures around Miyazono, Mashiki Town (Suzuki et al. 2018). The location of the map is shown in Fig. 17.1

epicenter of the Kumamoto earthquake mainshock was further west; however, the slip originated at 12 km depth beneath the epicenter and followed a complex propagation path. According to the rupture propagation process modeled by Kubo et al. (2016), slip with a relatively large displacement exceeding 1 m first appeared on the ground surface along the Futagawa fault near the boundary between Mashiki and Nishihara, 10 s after the onset of the mainshock. Then, the surface faults bilaterally spread eastward and westward from this point. As a result, the rupture propagated from east to west in Mashiki Town, consistent with the observed branching of the surface fault.

The displacement in the eastern part was ~60 cm and almost constant (Fig. 17.9). In the urban areas, the amount of displacement ranged from 5 to 40 cm, which was the minimum amount of displacement measured by the field survey.

To estimate fault displacement along the seismic fault at depth, it is necessary to combine the amounts of displacement along each fault branch. The total displacement was 60 cm in eastern Terasako, 50–60 cm in the vicinity of Kiyama and Miyazono, and did not decrease toward the west. Even in Yasunaga, the total amount of displacement reached 70 cm near Locs. KS11a and KS11b. Severe liquefaction occurred in the alluvial lowland of the Tessa River, making it challenging to confirm rupture on the ground surface to the west. The possibility that deep faults continue further to the west cannot be ignored.

Fig. 17.4 Photographs of the surface ruptures around Terasako, Mashiki Town (Locs. T01–T24) (compiled from Suzuki et al. 2018). White arrows indicate the surface ruptures. The directions toward which the photos were taken are shown in the parentheses. The location of each place is shown in Fig. 17.2

Fig. 17.5 Photographs of the surface ruptures around Terasako, Mashiki Town (Locs. T26a–T38) (compiled from Suzuki et al. 2018). White arrows indicate the surface ruptures. The directions toward which the photos were taken are shown in the parentheses. The location of each place is shown in Fig. 17.2

Fig. 17.6 Photographs of the surface ruptures along the southern fault trace in Mashiki (Locs. KS03–KS12) (compiled from Suzuki et al. 2018). White arrows indicate the surface ruptures. The directions toward which the photos were taken are shown in the parentheses. The location of each place is shown in Fig. 17.3

Fig. 17.7 Photographs of the surface ruptures along the central fault trace in Mashiki (Locs. KC04–KC06) (compiled from Suzuki et al. 2018). White arrows indicate the surface ruptures. The directions toward which the photos were taken are shown in the parentheses. The location of each place is shown in Fig. 17.3

Fig. 17.8 Photographs of the surface ruptures along the northern fault trace in Mashiki (Locs. KN07–KN15) (compiled from Suzuki et al. 2018). White arrows indicate the surface ruptures. The directions toward which the photos were taken are shown in the parentheses. The location of each place is shown in Fig. 17.3

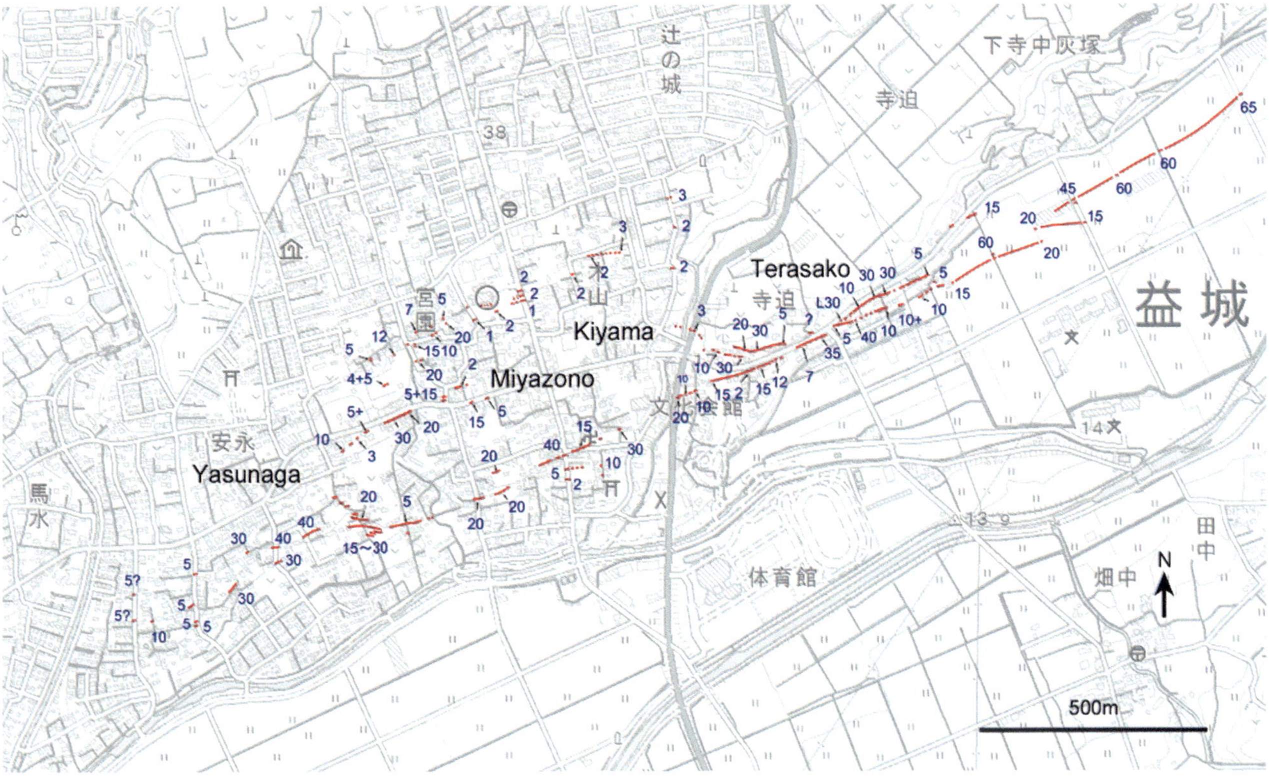

Fig. 17.9 Distribution of right-lateral displacement on the surface ruptures in Mashiki Town (Suzuki et al. 2018)

17.4 Relationship Between Tectonic Landforms and Earthquake Faults

Before the Kumamoto earthquake sequence, active faults had not been identified along the earthquake fault in Mashiki Town. A detailed survey after the earthquake revealed tectonic landforms in this area (Hideaki Goto et al. 2017). We performed a detailed interpretation of a 2-m-grid DEM anaglyph image and aerial photographs of this area and created a corresponding geomorphological map (Fig. 17.10). Kiyama and Miyazono are located on the Aso-4 pyroclastic flow deposit and its slightly younger eroded terrace (M surface). According to a boring survey by MLIT (2017), the Aso-4 pyroclastic flow deposits were confirmed at a depth of at least 10–15 m, and a south-facing flexure scarp and northward back-tilting were identified on the M surface. Fault scarps several meters high were also confirmed.

There is a clear flexure scarp on the north side of the prefectural road 28 ("Route 28" in Fig. 17.10). Previous research (e.g., Kumamoto Prefecture 1979; MLIT 2017) described this flexure scarp as an erosional scarp; however, the compositional layer is the same on the north and south of this scarp. On the other hand, northward back tilting is recognized on the M surface on the south side of the prefectural road 28, forming a small graben along the prefectural road with the flexure scarp. We speculate that this road was constructed along this graben.

At Miyazono, the dissected valley on the M surface bends 100 m to the right (point A in Fig. 17.10). Moreover, the L1 surface formed by the Sako-gawa River and Akitsu-gawa River is observed in Kiyama. Along the earthquake fault in Kiyama, the dissected valley on the L1 surface bends 60 m to the right (point B in Fig. 17.10). Although the amount of bending is not always the same as the amount of lateral displacement, the lateral displacement is interpreted to be larger than the vertical displacement on this fault because the amount of flexural displacement on the M surface is no more than 10 m, even when the displacements along the two traces are combined.

The following features are confirmed by comparing the tectonic landforms and earthquake faults:

(a) At Terasako, there is a clear relationship between preexisting grabens and 2016 earthquake faults.
(b) At Kiyama, the earthquake fault was confirmed on the top of the preexisting flexure scarp, and fault scarp topography is also observed on its western extension.

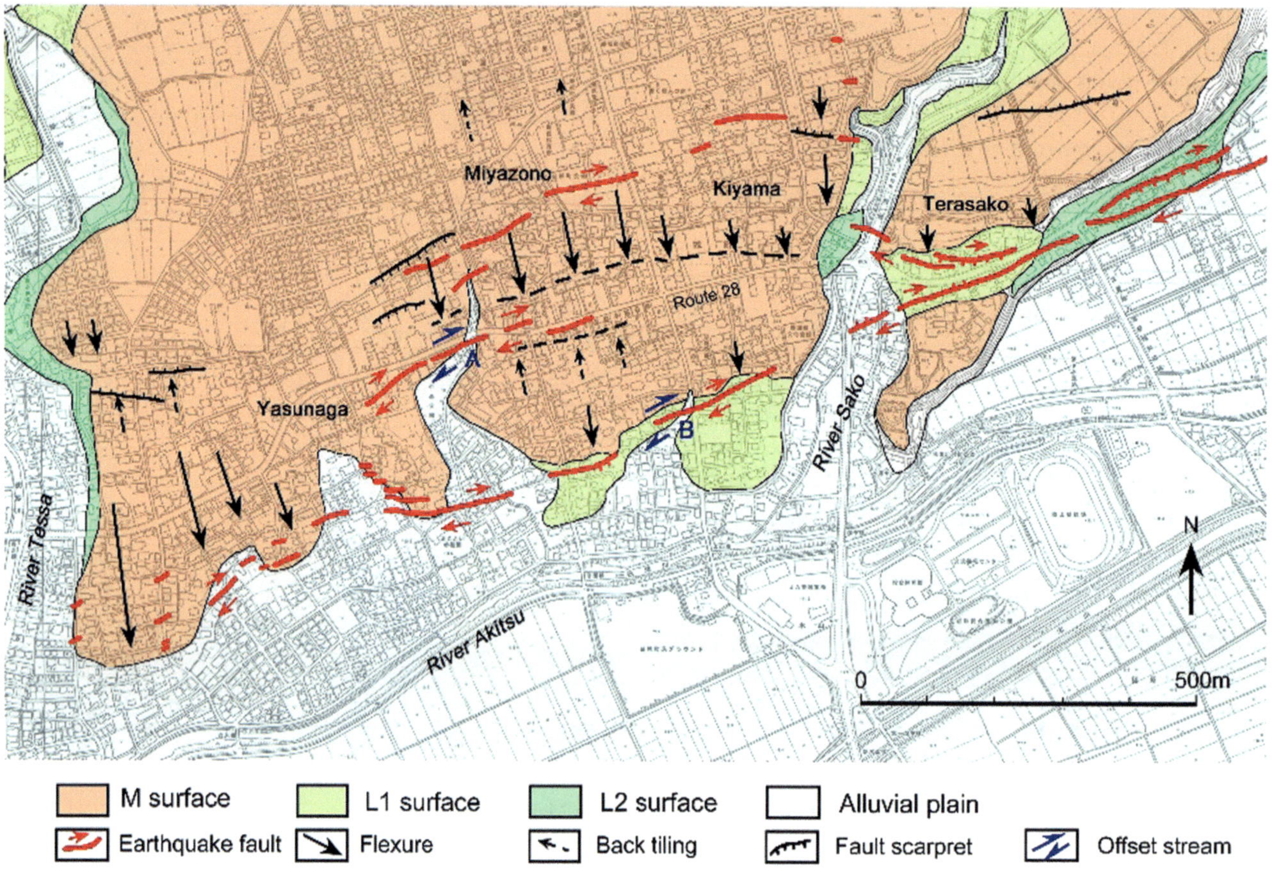

Fig. 17.10 Tectonic landform and the 2016 earthquake fault at the center of Mashiki Town (Suzuki et al. 2018)

(c) A preexisting graben is also recognized along the prefectural road 28 from Miyazono to Yasunaga, and 2016 earthquake faults appeared along the southern margin of the graben.

(d) The southern trace of the earthquake fault was located at the base of the flexure scarp on the southern edge of the M surface and continued for more than 1 km from Kiyama to Yasunaga. There was no evident vertical displacement during the earthquake.

(e) Large deflections of the valleys are recognized along the 2016 earthquake fault. The total deflection at points A and B reaches 160 m, and the average slip rate is calculated as 1.8 mm/yr or more, assuming that the M surface is 90,000 years old. If the same lateral displacement that occurred during the Kumamoto earthquake (70 cm) is repeated, the recurrence interval is calculated to be less than 1000 years. In contrast, Hideaki Goto et al. (2017) suggested that this fault may cause larger displacement in the future with a much longer interval.

17.5 A Preliminary Study on the Relationship Between Building Damage and Earthquake Faults

Figure 17.11 shows the distribution of damaged buildings by the mainshock, which was analyzed using aerial photographs (Monma et al. 2017), and the trace and displacement of the earthquake faults reported by Suzuki et al. (2018). In this figure, the damage is shown by red (rank 4), blue (rank 3), and yellow-green (rank 2) colors, with higher rank indicating severer damage.

In the Terasako area, many buildings with rank 4 damage are observed near the earthquake fault, and there are almost no undamaged buildings. In Kiyama and Miyazono districts, rank 4 and rank 3 damage is distributed immediately above the earthquake fault, and the number of buildings without damage is minimal. Rank 4 damage is more common between the southern and middle traces and south of the northern trace.

Figure 17.12 shows the north–south cross-sections near Terasako, Kiyama, and Yasunaga and illustrates the location of earthquake faults and the distribution of building damage. In the vicinity of Terasako (Fig. 17.12a), rank 4 and 3 damage is concentrated along the earthquake faults located on both sides of the prefectural road 28. In the north–south section near Kiyama (Fig. 17.12b), rank 4 damage is concentrated between the southern trace (Locs. TS01 to TS04) and the northern trace (Locs. TN04 to TN07). Building damage due to liquefaction is significant south of the southern trace (along the Akitsu-gawa River). In the vicinity of Yasunaga (Fig. 17.12c), there is substantial rank 2 damage on the alluvial lowlands along the Akitsu-gawa River; however, there are large amounts of rank 4 damage between the southern and northern traces.

The following preliminary conclusions are drawn from this analysis. (1) Structural damage distribution in the urbanized area of Mashiki Town indicated that displacement of the earthquake fault might have caused the building damage. (2) Building damage was severe in areas where the displacement exceeded 10 cm.

When discussing the relationship between earthquake faults and building damage immediately above the earthquake fault, it is not easy to separate the effects of ground motion and displacement. It is undeniable that both surface earthquake faults and strong ground motion caused building damage in Mashiki Town. Therefore, this research questions the validity of the claims made by MLIT (2017) that displacement along earthquake faults did not cause significant damage to low-rise buildings.

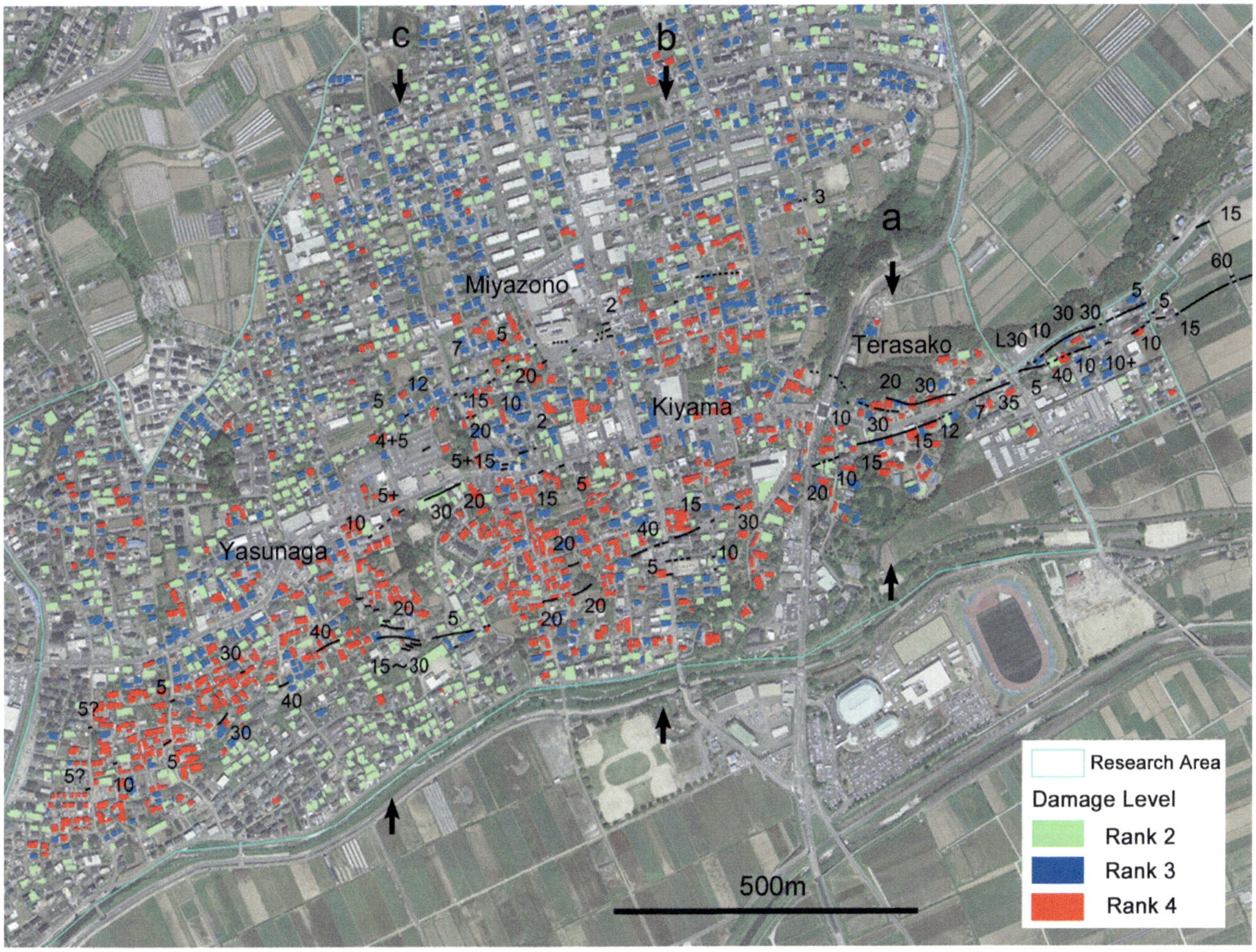

Fig. 17.11 Distribution of damaged buildings by the mainshock (Monma et al. 2017) and the earthquake faults (Suzuki et al. 2018). The damage is shown by red (rank 4), blue (rank 3), and yellow-green (rank 2) colors, with higher rank indicating severer damage. Black lines indicate the earthquake faults

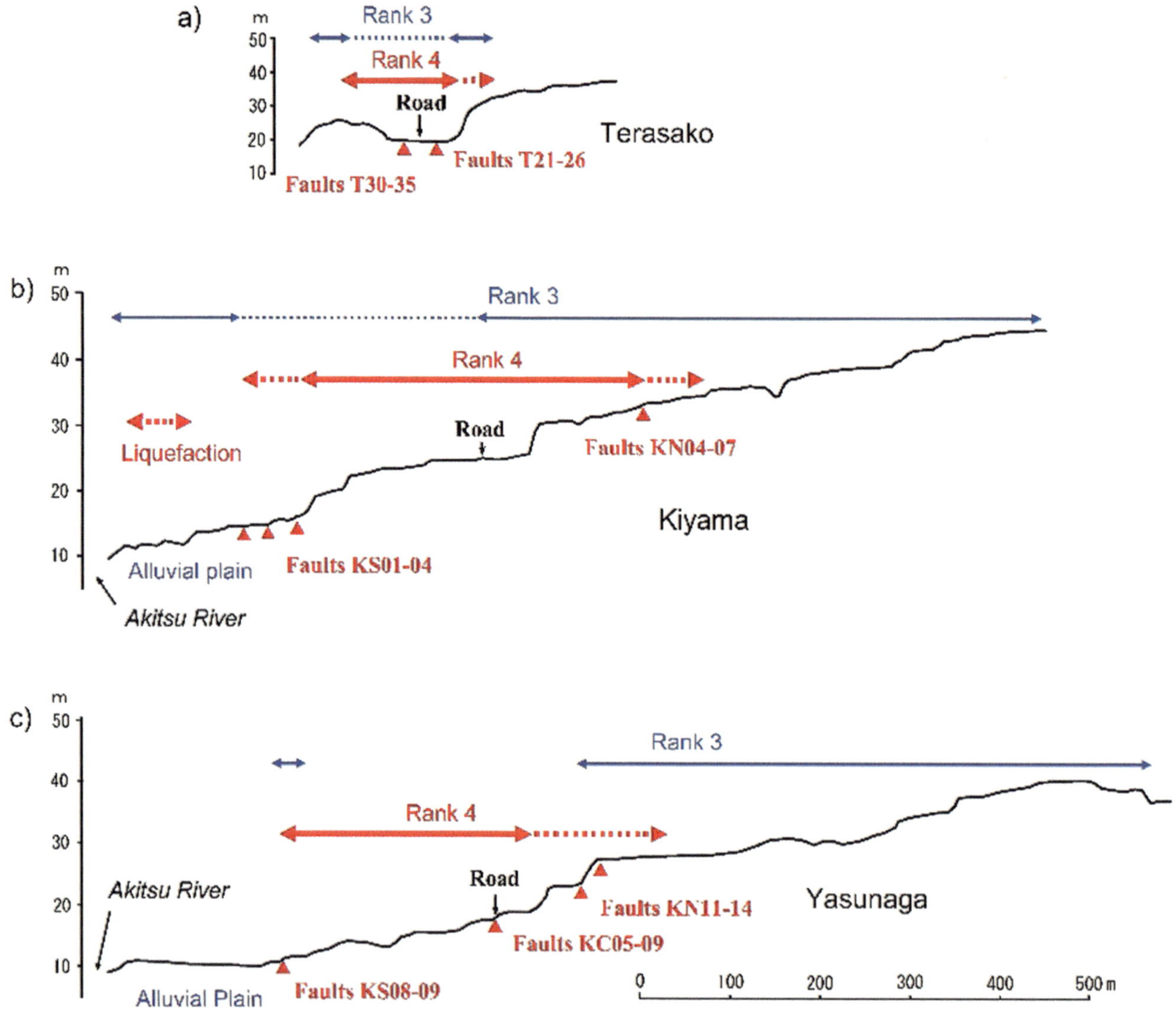

Fig. 17.12 Cross-sections in the north–south direction showing the relation between earthquake faults and building damages. Building damages were classified into 4 ranks by Monma et al. (2017) (Suzuki et al. 2018). **a** Terasako, **b** Kiyama, **c** Miyazono-Yasunaga. The location of each profile is shown in Fig. 17.11

Acknowledgements This work was supported by the Japan Society for the Promotion of Science (JSPS), Grants-in-Aid for Scientific Research (KAKENHI) (Grant Numbers: 18H03601; Principal Investigator: Yasuhiro Suzuki, and 16H06298; Principal Investigator: Hiroshi Shimizu).

References

Goto Hideaki, Tsutsumi H, Toda S, Kumahara Y (2017) Geomorphic features of surface ruptures associated with the 2016 Kumamoto earthquake in and around the downtown of Kumamoto City, and implications on triggered slip along active faults. Earth Planets Space 69:26–37

Goto Hiroyuki, Hata Y, Yoshimi M, Yoshida N (2017) Nonlinear site response at KiK-net KMMH16 (Mashiki) and heavily damaged sites during the 2016 Mw 7.1 Kumamoto Earthquake, Japan. Bull Seismol Soc Am 107:1802–1816. https://doi.org/10.1785/0120160312

Kubo H, Suzuki W, Aoi S, Sekiguchi H (2016) Source rupture processes of the 2016 Kumamoto, Japan, earthquakes estimated from strong-motion waveforms. Earth Planets Space 68:161–173

Kumahara Y, Okada S, Kagohara K, Kaneda H, Goto H, Tsutsumi H (2017) Active fault map 'Kumamoto" (revised edition). Geospatial Information Authority of Japan D1-No.868, Geospatial Information Authority of Japan, Tokyo

Kumamoto Prefecture (1979) 1:50,000 landform classification map "Mifune". https://nlftp.mlit.go.jp/kokjo/tochimizu/F3/data/L/4306L.jpg. Accessed 27 July 2021

Ministry of Land, Infrastructure, Transport and Tourism (2017) Report of the expert's conference on safety countermeasures towards reconstruction of the city area of Mashiki Town. http://www.mlit.go.jp/report/press/toshi08_hh_000034.html. Accessed 27 July 2021

Monma N, Fujiwara H, Nakamura H, Saeki T, Naito S, Shimomura H, Yamada T (2017) Distribution of collapsed buildings near the surface deformation associated with the Kumamoto earthquake. In: Proceedings of Japan Association for Earthquake Engineering, University of Tokyo, Tokyo, 13–14 Nov 2017

Nakata T, Shimazaki K, Suzuki Y, Tsukuda E (1998) Fault branching and directivity of rupture propagation. J Geogr 107:512–528

Nakata T, Yomogida K, Odaka J, Sakamoto T, Asahi K, Chida N (1995) Surface fault ruptures associated with the 1995 Hyogoken-Nanbu Earthquake. J Geogr 104:127–142

Suzuki Y, Watanabe M (2006) Logic for recognizing surface fault rupture associated with the 2006 Niigataken Chuetsu Earthquake. Chikyu Special Issue 53:90–98

Suzuki Y, Watanabe M, Nakata T (2016) Lessons learned from the 2016 Kumamoto Earthquake and suggestion toward active fault prevention. Kagaku 86:839–847

Suzuki Y, Watanabe M, Nakata T (2018) Surface earthquake faults in the urbanized area of Mashiki town associated with the 2016 Kumamoto Earthquake of Japan—Tectonic significance and impact on building damages. Active Fault Res 48:13–34

Watanabe M, Suzuki Y, Nakata T (2016) Earthquake fault and "severely damaged zone" in Mashiki Town and Minami-Aso Village. Japan Geoscience Union Meeting 2016, MIS24-P90

Surface Ruptures Accompanied with the Largest Foreshock

18

Nobuhiko Sugito, Hideaki Goto, Yasuhiro Kumahara, Hiroyuki Tsutsumi, Takashi Nakata, Kyoko Kagohara, Nobuhisa Matta, and Mitsuhisa Watanabe

Abstract

The largest foreshock (Mj 6.5, Mw 6.2) of 21:26 (JST), April 14, 2016, caused surface fault ruptures in the Takaki–Fukuhara area (including the Itera–Oike area), and the ruptures enlarged associated with the mainshock (Mj 7.3, Mw 7.0) of 01:25 (JST), April 16, 2016, as reported by previous studies. This indicates that the same fault slipped repeatedly during the largest foreshock and mainshock. We conducted additional investigations of surface ruptures related to the largest foreshock, and newly identified surface ruptures accompanied with the largest foreshock that became larger associated with the mainshock in Takigawa, Takaki, and other areas. As a result, the length of the surface ruptures accompanied with the largest foreshock is at least 7.5 km, as compared to 6 km reported by the previous study. In this chapter, we describe the surface ruptures from south to north, including those reported by previous studies, in order to understand the foreshock–mainshock sequence of the 2016 Kumamoto earthquake.

Keywords

Largest foreshock • Right-lateral strike-slip offset • Repeated slip • Foreshock–mainshock sequence

The largest foreshock (M_j 6.5, Mw 6.2) occurred in the southern part of the mainshock-source area at 21:26 (JST), April 14, 2016, ~28 h before the mainshock. Sugito et al. (2016) reported that (1) the largest foreshock produced approximately 6-km-long surface ruptures from Takaki to Fukuhara (including the Itera–Oike area), (2) these ruptures became larger associated with the mainshock, and (3) the foreshock–mainshock sequence was not successive events on the adjacent different fault zones. Here, we reexamine

Supplementary Information
The online version contains supplementary material available at (https://doi.org/10.1007/978-981-19-1150-7_18).
Note
The supplementary material of this book contains geographical information system (GIS) and spreadsheet files of the surface ruptures associated with the mainshock of the 2016 Kumamoto earthquake sequence

N. Sugito (✉)
Faculty of Sustainability Studies, Hosei University, 2-17-1, Fujimi, Chiyoda-ku, Tokyo, 102-8160, Japan
e-mail: nsugito@hosei.ac.jp

H. Goto
Graduate School of Humanities and Social Sciences, Hiroshima University, 1-2-3, Kagamiyama, Higashi-Hiroshima, Hiroshima, 739-8522, Japan
e-mail: hgoto@hiroshima-u.ac.jp

Y. Kumahara
Graduate School of Humanities and Social Sciences, Hiroshima University, 1-1-1, Kagamiyama, Higashi-Hiroshima, Hiroshima, 739-8524, Japan
e-mail: kumakuma@hiroshima-u.ac.jp

H. Tsutsumi
Department of Environmental Systems Science, Faculty of Science and Engineering, Doshisha University, 1-3, Tatara-Miyakodani, Kyotanabe, Kyoto, 610-0394, Japan
e-mail: htsutsum@mail.doshisha.ac.jp

T. Nakata
Department of Geography, Hiroshima University, 1-2-3, Kagamiyama, Higashi-Hiroshima, Hiroshima, 739-8522, Japan
e-mail: tnakata@hiroshima-u.ac.jp

K. Kagohara
Faculty of Education, Yamaguchi University, 1677-1, Yoshida, Yamaguchi, Yamaguchi, 753-8513, Japan
e-mail: k-kago@yamaguchi-u.ac.jp

N. Matta
Graduate School of Education, Okayama University, 3-1-1, Tsushima-naka, Kita-ku, Okayama, Okayama, 700-8530, Japan
e-mail: matta@okayama-u.ac.jp

M. Watanabe
Faculty of Sociology, Toyo University, 5-28-20, Hakusan, Bunkyo-ku, Tokyo, 112-8606, Japan
e-mail: manqq@toyo.jp

Y. Kumahara et al. (eds.), *Surface Ruptures Associated with the 2016 Kumamoto Earthquake Sequence in Southwest Japan*, Advances in Geological Science, https://doi.org/10.1007/978-981-19-1150-7_18

surface ruptures accompanied with the largest foreshock and report some new discoveries after publication of Sugito et al. (2016).

As Sugito et al. (2016) described, we conducted field observations on April 15 (after the largest foreshock and before the mainshock) and after the mainshock. Eyewitness accounts also enabled us to identify surface ruptures associated with the largest foreshock. Besides, we mapped surface ruptures by interpreting high-resolution aerial photographs taken on April 15 by the Geospatial Information Authority of Japan (GSI).

We also mapped surface ruptures caused by the earthquake sequence using high-resolution aerial photographs taken on April 16 and 20 (immediately after the mainshock) by GSI. As a result, part of the surface ruptures identified in the field could not be mapped by the high-resolution aerial

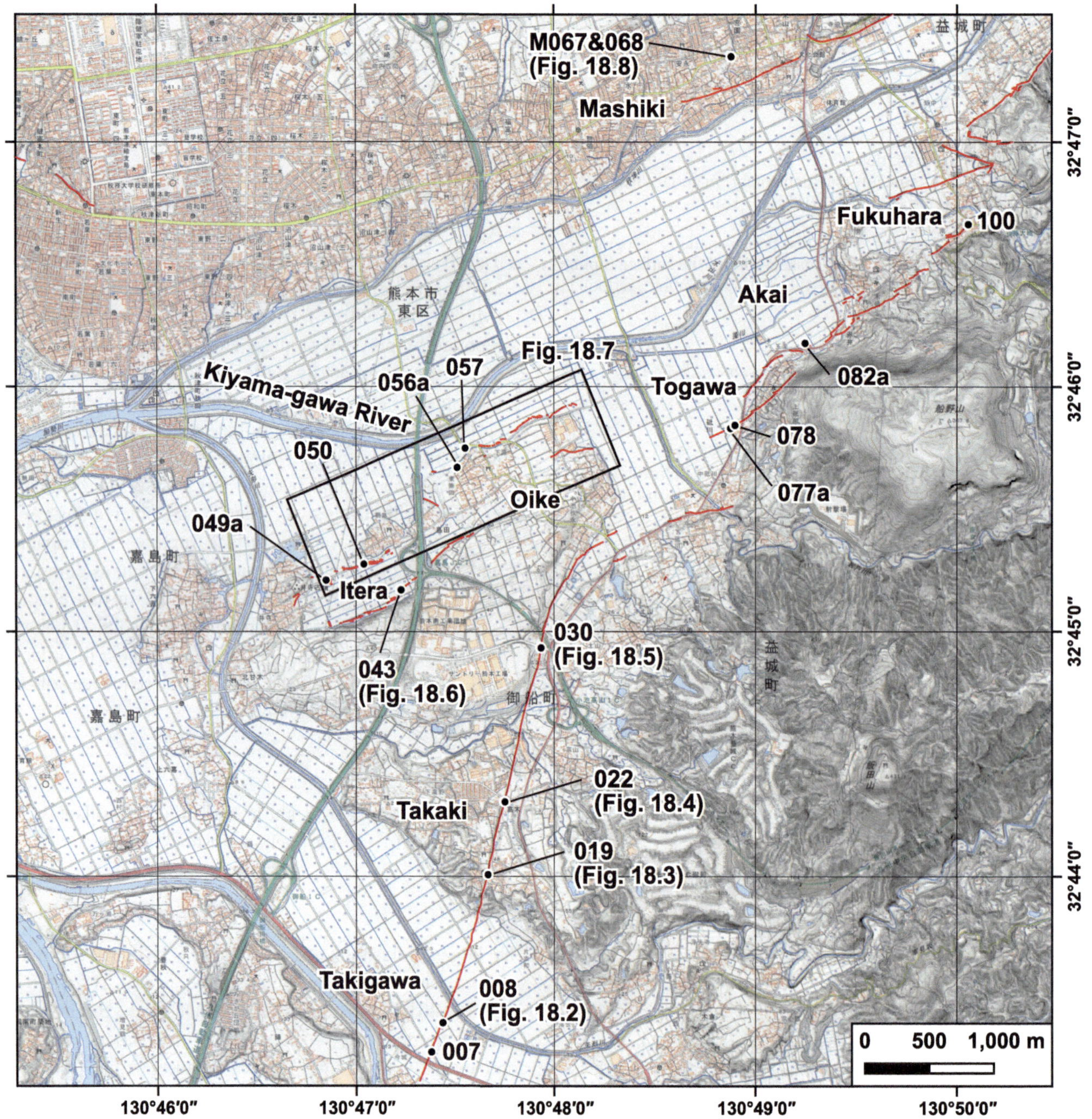

Fig. 18.1 The observation sites of the surface ruptures accompanied with the largest foreshock. The red line is the trace of the surface ruptures of the 2016 Kumamoto earthquake mainshock. The background is a half-transparent topographic map of Geospatial Information Authority of Japan (GSI) underlain by a slope map produced from GSI 5-m grid digital elevation models (DEMs)

photographs. This indicates that we may have missed part of the surface ruptures accompanied with the largest foreshock when interpreting high-resolution aerial photographs of April 15. Additional interviews with local people may partly solve this problem.

We newly identified a slight right-lateral strike-slip offset of vegetable fields at Loc. 008 in Takigawa (32° 43′ 24.05″ N, 130° 47′ 26.23″ E) (Fig. 18.1), based on high-resolution aerial photographs of April 15 (Fig. 18.2). The surface color east of the surface rupture trace seems darker than that west of the trace. This surface rupture enlarged associated with the mainshock, as indicated by Fig. 18.2b. The right-lateral strike-slip offset after the mainshock was measured at ~35 cm in the field, with no vertical displacement. In addition, we could identify a fracture line that runs across a paved road at Loc. 007 (32° 43′ 16.92″ N, 130° 47′ 22.87″ E) (Fig. 18.1), based on high-resolution aerial photographs of April 15. This fracture line probably corresponds to the surface rupture caused by the largest foreshock.

For the site immediately south of Loc. 019 in Takaki (32° 43′ 59.60″ N, 130° 47′ 39.77″ E) (Fig. 18.1), we newly obtained evidence for repeated right-lateral strike-slip associated with the earthquake sequence (Fig. 18.3). Here, the U-shaped drainage ditch became deflected right-laterally due to the largest foreshock (Fig. 18.3b), and the deflection enlarged associated with the mainshock (Fig. 18.3a). Also, left-stepping cracks emerged during the largest foreshock (Fig. 18.3d) and became more extensive due to the mainshock (Fig. 18.3c). Local personnel provided us these photographs on July 20, 2019, and almost all of these surface ruptures had disappeared by restoration by that day. Thus, we tried to measure right-lateral strike-slip offset amounts caused by the largest foreshock and mainshock using these photographs, with reference to the width of the U-shaped ditch that partly remained until that day. As a result, offset amounts were estimated as ~6 cm and ~34 cm for the largest foreshock and mainshock, respectively. The resultant estimation of ~40 cm is consistent with the offset amounts measured at Locs. 018 and 019 after the mainshock (~50 cm and ~60 cm, respectively). Therefore, we conclude that the right-lateral strike-slip offset during the mainshock (~34–54 cm) was ~6–9 times that associated with the largest foreshock, which is concordant with the earthquake magnitude of each event.

The surface rupture at Loc. 022 in Takaki (32° 44′ 18.25″ N, 130° 47′ 45.12″ E) (Fig. 18.1) has already been reported by Sugito et al. (2016). We visited here on April 15 and found surface ruptures on a paved road (Fig. 18.4), and the mainshock enlarged these ruptures. We measured a right-lateral

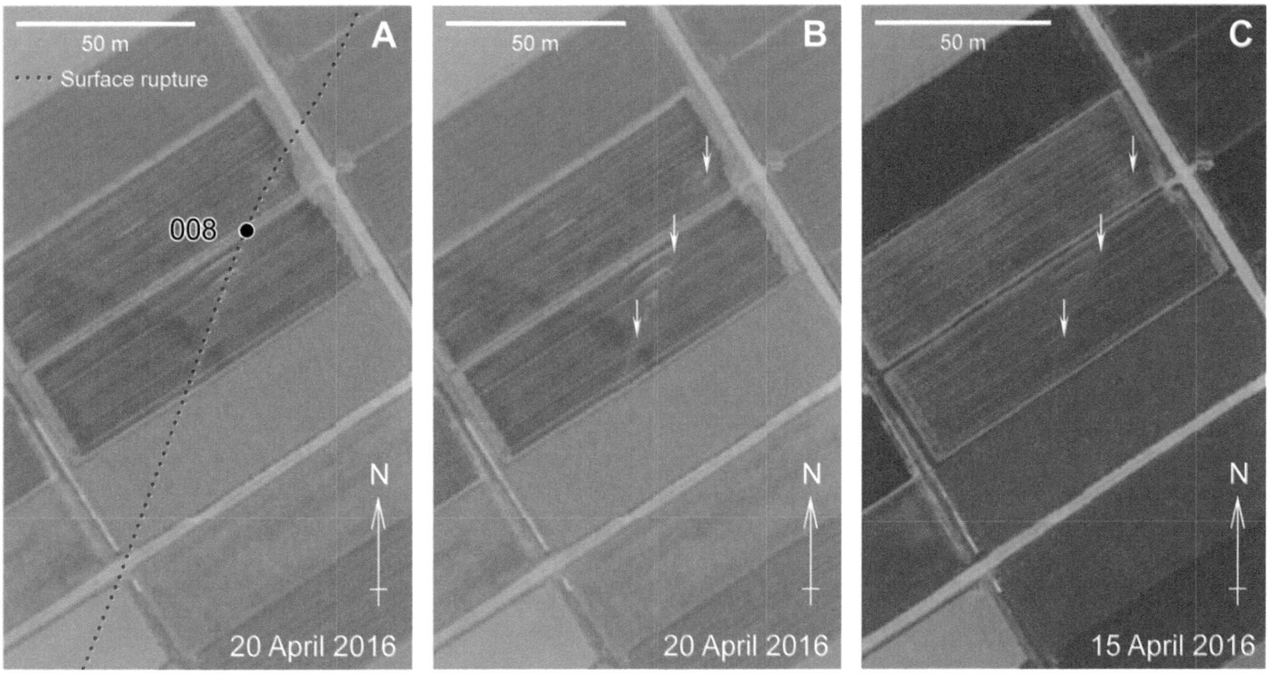

Fig. 18.2 Surface ruptures at Loc. 008, based on interpretation of high-resolution aerial photographs. See Fig. 18.1 for the location. **a** and **b** are post-mainshock orthophoto images, and **c** is a pre-mainshock orthophoto image (after the largest foreshock). A slight right-lateral strike-slip offset is identified in vegetable fields as indicated by white arrows in **c**, and this offset became larger associated with the mainshock as shown in **b**. The final offset amount was measured at ~35 cm in the field. The base aerial photographs are from the GSI Maps (Geospatial Information Authority of Japan 2021)

Fig. 18.3 Photographs of the surface ruptures at the site immediately south of Loc. 019. See Fig. 18.1 for the location. **a** and **c** were taken after the mainshock. **b** and **d** were taken before the mainshock (after the largest foreshock). We can identify right-lateral strike-slip and left-stepping cracks caused by the largest foreshock that became more substantial associated with the mainshock. These photographs were provided to us by local personnel. Considering the width of the U-shaped ditch that partly remained until the day of our field observation, we estimated the offset amounts caused by the largest foreshock and mainshock to be ~6 cm and ~34 cm, respectively, using these photographs

strike-slip offset amount of ~30 cm in the field, with no vertical separation. Also, we newly obtained an eyewitness account for Loc. 030 (32° 44′ 55.95″ N, 130° 47′ 56.08″ E) in Takaki. Here, a right-lateral strike-slip offset of ~55 cm was identified in a paved road after the mainshock, as well as an elongated ground fissure ~10 m east of and parallel to the right-lateral strike-slip trace (32° 44′ 55.83″ N, 130° 47′ 56.39″ E) (Fig. 18.5). This fissure appeared during the largest foreshock, with a several-cm right-lateral strike-slip offset, based on the eyewitness account. This fissure and right-lateral strike-slip offset enlarged associated with the mainshock, as indicated in Fig. 18.5.

The surface ruptures at the following four sites are described by Sugito et al. (2016): (1) Loc. 077a (32° 45′ 49.80″ N, 130° 48′ 52.52″ E) in Togawa, (2) Loc. 078 (32° 45′ 50.59″ N, 130° 48′ 53.92″ E) in Togawa, (3) Loc. 082a (32° 46′ 10.70″ N, 130° 49′ 14.69″ E) in Akai, and (4) Loc. 100 (32° 46′ 39.85″ N, 130° 50′ 03.40″ E) in Fukuhara (Fig. 18.1).

At Loc. 043 in Itera (32° 45′ 10.21″ N, 130° 47′ 13.80′ ′ E) (Fig. 18.1), we newly found open cracks due to the largest foreshock in paddy fields by interpreting high-resolution aerial photographs of April 15 (Fig. 18.6). During the mainshock, these cracks enlarged, and more cracks appeared, as indicated in Fig. 18.6. We measured a final right-lateral strike-slip offset of ~15 cm in the field, with the north side down vertical separation.

From Itera to Oike (Fig. 18.1), numerous surface ruptures were identified after the mainshock, part of which were detected after the largest foreshock (before the mainshock), based on high-resolution aerial photographs (Sugito et al. 2016) (Fig. 18.7). For this area, Sugito et al. (2016) describe the surface ruptures at the following four sites in detail: (1) Loc. 049a (32° 45′ 12.62″ N, 130° 46′ 51.05″ E), (2) Loc. 50 (32° 45′ 16.55″ N, 130° 47′ 02.40″ E), (3) Loc. 056a (32° 45′ 40.30″ N, 130° 47′ 30.64″ E), and (4) Loc. 057 (32° 45′ 44.96″ N, 130° 47′ 33.04″ E).

As a result of the above reexaminations, we reconfirm that the same portion of the fault ruptured to the surface twice during the Kumamoto earthquake sequence (Sugito

Fig. 18.4 Photographs of the surface ruptures at Loc. 022. Modified from Sugito et al. (2016). See Fig. 18.1 for the location. **a** and **b** were taken after the largest foreshock (before the mainshock). We can identify buckling and left-stepping cracks in a paved road. **c**–**e** were taken after the mainshock. We measured a right-lateral strike-slip offset amount of ~30 cm in the field after the mainshock, as shown in **e**

Fig. 18.5 Photographs of the surface ruptures at the site ~10 m east of Loc. 030. See Fig. 18.1 for the location. These photographs were provided to us by local personnel. **a** and **b** show that a fissure with a several-cm right-lateral strike-slip offset emerged associated with the largest foreshock. This fissure and right-lateral strike-slip offset became more extensive during the mainshock, as indicated by **c** and **d**

et al. 2016). However, the length of the surface ruptures accompanied with the largest foreshock should be revised from 6 km (Sugito et al. 2016) to at least 7.5 km because we newly identified the surface rupture accompanied with the largest foreshock at Locs. 007 and 008 in Takigawa.

We also examined whether the faults in the central Mashiki Town (Fig. 18.1) slipped during the largest foreshock, in addition to the mainshock. We identified surface break lines caused by the largest foreshock in a paved parking space between Locs. M067 and M068 (32° 47′ 20.93″ N, 130° 48′ 52.67″ E) in Miyazono, based on high-resolution aerial photographs of April 15 (Fig. 18.8). Locs. M067 and 068 correspond to Locs. KC05 and 06 in Suzuki et al. (2018), respectively. Moreover, more break lines were identified on high-resolution aerial photographs of April 16. We obtained an eyewitness account for the break line shown by the symbol "X" and its western extension (along the northern edge of a house) (Fig. 18.8). This line existed before the earthquake sequence as a pavement boundary. However, ground fissures emerged along this break line during the largest foreshock. Besides, field observation detected that right-lateral strike-slip and southern-side-down vertical displacement appeared associated with the mainshock. Because right-lateral strike-slip

Fig. 18.6 Surface ruptures at Loc. 043, based on interpretation of high-resolution aerial photographs. See Fig. 18.1 for the location. **a** and **b** are post-mainshock orthophoto images, and **c** is a pre-mainshock orthophoto image (after the largest foreshock). We can find numerous ground fissures in paddy fields in **b** as indicated by white arrows, part of which can also be identified in **c**. The base aerial photographs are from the GSI Maps (Geospatial Information Authority of Japan 2021)

was also identified after the mainshock at Locs. M067 and M068 (Suzuki et al. 2018), we imply that the right-lateral strike-slip fault here repeatedly slipped during the largest foreshock and mainshock. Further investigations are needed to examine what occurred during the largest foreshock in more detail.

Acknowledgements We thank the Geospatial Information Authority of Japan for providing us with the high-resolution aerial photographs. We are also grateful to the residents who provided us with eyewitness accounts and photographs for the surface ruptures.

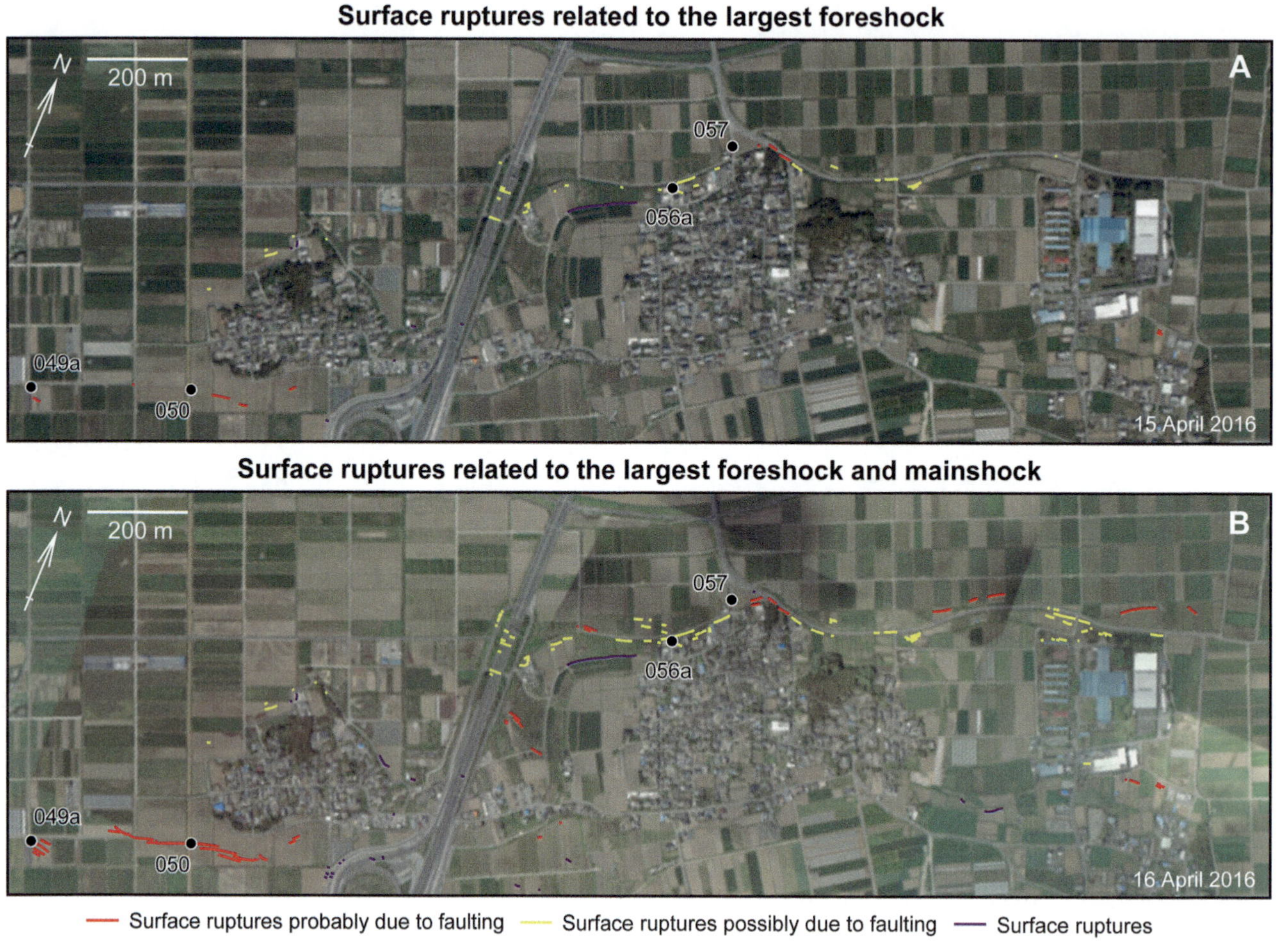

Fig. 18.7 Surface ruptures from Itera to Oike, based on interpretation of high-resolution aerial photographs. Modified from Sugito et al. (2016). See Fig. 18.1 for the location. **a** shows surface ruptures caused by the largest foreshock, and **b** shows surface ruptures caused by the largest foreshock and mainshock. We detected numerous surface ruptures after the mainshock, part of which emerged due to the largest foreshock. The base aerial photographs are from the GSI Maps (Geospatial Information Authority of Japan 2021)

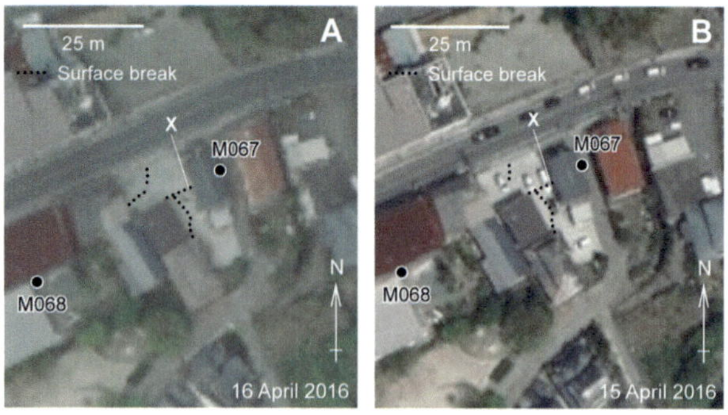

Fig. 18.8 Surface break at the site between Locs. M067 and M068, based on high-resolution aerial photographs. See Fig. 18.1 for the location. **a** is a post-mainshock orthophoto image, and **b** is a pre-mainshock orthophoto image (after the largest foreshock), with the mark of surface break shown. We identified more break lines after the mainshock than after the largest foreshock (before the mainshock). The break line shown by the symbol "X" and its western extension (along the northern edge of a house) existed before the earthquake sequence as a pavement boundary, along which ground fissures emerged due to the largest foreshock. Then, right-lateral strike-slip and south-side-down vertical separation were detected after the mainshock along this break line. The base aerial photographs are from the GSI Maps (Geospatial Information Authority of Japan 2021)

References

Geospatial Information Authority of Japan (2021) GSI Maps, https://maps.gsi.go.jp. Accessed 13 Oct 2021

Sugito N, Goto H, Kumahara Y, Tsutsumi H, Nakata T, Kagohara, K, Matsuta N, Yoshida H (2016) Surface fault ruptures associated with the 14 April foreshock (Mj 6.5) of the 2016 Kumamoto earthquake sequence, southwest Japan. Earth Planets Space 68:170. https://doi.org/10.1186/s40623-016-0547-5

Suzuki Y, Watanabe M, Nakata T (2018) Surface earthquake faults in the urbanized area of Mashiki town associated with the 2016 Kumamoto Earthquake of Japan: Tectonic significance and impact on building damages. Active Fault Res 48:13–34. https://doi.org/10.11462/afr.2018.48_13